The Minerals, Metals & Materials Series

More information about this series is available at http://www.springer.com/series/15240

Wu Xianxi

Inert Anodes for Aluminum Electrolysis

TMS

Springer

Wu Xianxi
Guizhou University
Guiyang, Guizhou, China

ISSN 2367-1181 ISSN 2367-1696 (electronic)
The Minerals, Metals & Materials Series
ISBN 978-3-030-28912-6 ISBN 978-3-030-28913-3 (eBook)
https://doi.org/10.1007/978-3-030-28913-3

This Springer imprint is published by the registered company Springer Nature Switzerland AG
The registered company address is: Gewerbestrasse 11, 6330 Cham, Switzerland

Preface

In 1886, Hall of the United States and Heroult of France patented aluminum smelting by electrolytic method; since then aluminum electrolytic production has developed rapidly. Primary aluminum is now the most important metal material, and its output is second only to iron and steel.

Aluminum production has developed rapidly, and carbon anode has been used in modern aluminum electrolysis production. However, the production of aluminum electrolysis has been in a state of high energy consumption in the past 100 years. Therefore, the anode must be replaced regularly during production. As a result, the cost is high and production is unstable. In addition, because aluminum electrolysis is carried out in cryolite alumina melt containing fluorine, a large amount of greenhouse gases, carcinogenic carbon, and fluorine compounds are produced in the production process. This seriously pollutes the environment and aggravates the working conditions. Therefore, the electrolytic aluminum industry in various countries is looking for technological innovations that are energy-saving, environmentally friendly, and have higher economic benefits. Inert anode technology is believed to meet this requirement. Replacing the traditional carbon anode with inert anode can completely solve the problem of production of greenhouse gases, polycyclic aromatic hydrocarbons (PAHs), and carbon oxides and improve the environmental conditions. In addition, since this technology does not consume carbon anode, the investment cost will be reduced. If inert electrodes are used together with wet cathodes, it is expected to increase power efficiency by 25% and reduce operating costs by more than 10%.

Because the use of inert anode will cause the aforementioned revolutionary changes in the aluminum electrolysis industry, all countries are competing to conduct research. In recent years, aluminum electrolytic inert anode has been the frontier and hotspot of aluminum metallurgy research in the world.

According to the material composition, inert anode can be classified into three categories: ceramic anode, metal anode, and cermet anode. Based on the previous research, it is considered that the cermet inert anode (Cu-NiO/Fe_2O_3) of American Aluminum Company (Alcoa) is the best inert anode at present.

At the beginning of this century, Alcoa conducted 265 h of industrial testing on a 6000A electrolyzer. The experimental results show that the anodic corrosion rate is 1.8 cm/a, and the purity of aluminum is 99.53%. It is pointed out that cermet is suitable for aluminum electrolysis and is the most likely material to become an inert anode for aluminum electrolysis. However, the test results show some problems. The cermet has the characteristics of ceramic brittleness. Thus, the cermet anode is easy to break, the poor machinability is difficult to process connection hanging, it cannot meet large-scale demand, etc. This hinders its application in the electrolytic aluminum industry.

The emergence of nano-ceramics at the end of the twentieth century brings new hopes for solving the brittleness and poor machinability of ceramics. Nano-ceramics have many excellent properties, such as wear resistance, corrosion resistance, high temperature resistance, high hardness, resistance to aging, high toughness, superplastic, and so on. The superplastic property of nano-ceramics enables them to be forged, extruded, stretched, and bent; perform other special processing like other materials; and then directly prepare precision-sized parts. High toughness improves the brittleness of ceramics.

The author of this book added nanomaterials to the above cermet to prepare Cu-NiO/Fe_2O_3 nanocomposite cermet to solve the shortcomings of brittleness, difficult to large scale production and poor machinability of cermet. This can improve the technology of aluminum electrolysis.

The book is divided into three parts and a total of five chapters.

The first part includes the first and second chapters. This part briefly reviews the development of aluminum electrolysis, as well as the electrolysis cells and carbon anodes in the aluminum industry. In addition, the shortcomings of electrolytic cells and carbon anodes in the aluminum industry are introduced.

The second part introduces the inert anode. The inert anode is proposed to overcome the shortcomings of the existing electrolytic cells and carbon anode. The development, advantages, and disadvantages of various inert anodes and electrolytic cells are introduced. This part also focuses on the cermet inert anode developed by American Aluminum Company, which is considered to be the best and most promising inert anode. In addition, the shortcomings of cermet inert anode and the problems that hinder its popularization and application in the aluminum industry are introduced.

The third part covers nano-cermet and nano-cermet inert anode. In order to overcome the shortcomings of cermet inert anode, nano-cermet has been used as inert anode materials. The preparation, characterization, physical properties, conductivity, high temperature resistance, and high corrosion resistance of nano-cermet are introduced. The results show that nano-cermet has good machinability and thermal shock resistance. The nano-cermet can be used as inert anode for aluminum electrolysis.

Guiyang, Guizhou, China Wu Xianxi

Contents

Part I
The Development of Inert Anodes for Aluminum Electrolysis

Chapter 1
Research Background of Inert Anodes for Aluminum Electrolysis

1.1 The Development of Aluminum Metallurgy

As early as in the first century AD, the Roman Gaius Plinius wrote articles about aluminum. However, aluminum metallurgy first appeared in 1825. Since then, aluminum metallurgy has developed rapidly.

The first chemical process of aluminum smelting was in 1825, when the German F. Wohler reduced anhydrous aluminum chloride with potassium to produce aluminum. In 1845, the Frenchman H. S. Deville reduced the mixed salt $NaCl–AlCl_3$ with sodium to obtain aluminum. Then, Rossi and Beketov reduced cryolite to aluminum with sodium and magnesium, respectively. They established the smelter by this method. The total amount of aluminum produced by chemical methods was about 200 metric t.

In 1886, Hall of the United States and Heroult of France proposed a patent of aluminum production by using the cryolite and alumina molten salt electrolysis method, which improves the development of aluminum. Initially, they used a small prebaked electrolysis cell. At the beginning of the twentieth century, a small self-baking anode electrolysis cell (Soderberg) with current supply through horizontal anode studs appeared. The amperage of the cell gradually developed from the initial 2–50 kA or greater. In the 1940s, Soderberg cells with vertical anode studs appeared. After the 1950s, the emergence of prebaked anode batteries brought electrolytic aluminum smelting technology into a new stage of large-scale development and modernization. The modern aluminum industry has basically eliminated the self-baking anode aluminum electrolysis cell and now mainly uses large-scale prebaked anode aluminum electrolysis cell with amperages greater than 160 kA.

At present, the capacity of the electrolytic cells has developed from over 280 kA and up to 600 kA, as shown in Fig. 1.1. The design, installation, and operation control of electrolytic cells have been established on the basis of modern computer technology and artificial intelligence. The technical and economic indexes and the environmental protection level of electrolytic aluminum smelting have been greatly

W. Xianxi, *Inert Anodes for Aluminum Electrolysis*, The Minerals, Metals & Materials Series, https://doi.org/10.1007/978-3-030-28913-3_1

Fig. 1.1 600 kA aluminum reduction cell. (China Weiqiao Aluminum and Electric Co., Ltd.)

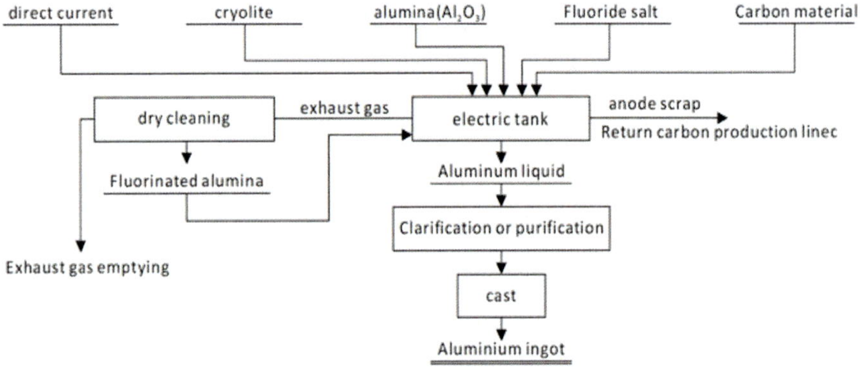

Fig. 1.2 Aluminum electrolysis process

improved. Although electrolytic aluminum smelting is still based on cryolite-alumina molten salt electrolysis, it has been made great progress both in theory and application, and it is developing.

The cryolite-alumina molten salt electrolysis process of Hall and Heroult mainly consists of two steps:

1. Production of raw materials, including alumina, fluoride salts, carbon materials, and other raw materials for electrolysis.
2. Metal aluminum electrolysis production. The process of modern electrolytic aluminum smelting is shown in Fig. 1.2.

1.2 Development of Aluminum Electrolytic Cells and Anodes

The electrolytic cell is the main equipment for aluminum production. Since Hall of the United States and Heroult of France discovered the cryolite-alumina molten salt electrolysis process in 1886, the production technology of aluminum electrolysis and the aluminum reduction cell has made many progress and improvements. Among these, the anode used in electrolysis has the maximum improvement in the past 133 years.

Prebaked anode electrolysis cells are classified into a discontinuous type (prebake) and a continuous type (Soderberg). The former has been most widely used in the aluminum industry.

The discontinuous prebaked anode electrolysis cells (hereinafter referred to as prebake cells) are the earliest type of aluminum reduction cell. In the first industrial period (1886–1900), the current of the cell was only between 4 and 8 kA, the current efficiency was low (70%), and the power consumption (42 kW h/kg Al) was very high. Thus, the cells were soon replaced by self-baking anode cells. In the mid-1950s, modern large prebaked anodized aluminum reduction cells began to appear for three reasons:

1. It is difficult to increase the anode size of the self-baking cell, which is not suitable for further increasing the amperage and production capacity.
2. The modern development of the electrode production industry can produce large prebaked anodes. The appearance of high-power and high-efficiency rectifiers is beneficial for increasing the current capacity of the electrolytic cells.
3. The prebaked anode cell is simple in structure and can save steel and other furnace materials. It can collect the anode gases effectively, which is beneficial for environmental protection.

1.3 Three Parts of Prebaked Cells

The prebaked cell consists of three parts: anode device, cathode device, and the conductive busbar system.

1.3.1 Anode Device

The anode device is composed of anode beam, anode carbon block group, and anode lifting mechanism.

The anode beam bears the weight of all anodes, and the current is fed into the electrolytic cell through the anodes. It is made of cast aluminum and can be moved up and down by a lifting mechanism to adjust the position of the anodes.

A prebaked cell has a plurality of anode carbon blocks. Each group includes one, two, or three prefabricated anode carbon blocks. The carbon block, the yoke, and the aluminum rod are assembled to form the electrolytic anode, as shown in Fig. 1.3.

1、Anode carbon block 2、Steel yoke 3、Aluminum rod 4、anode beam
5、Cathode carbon blocks 6、Cell hoods.

Fig. 1.3 Schematic diagram of a cross section of a modern aluminum reduction cell

The number of prebaked anode carbon blocks depends on the current intensity of the cell, the current density of the anode, and the geometry of the carbon anode block. For example, if the anode current density is about 0.7 A/cm^2 in a 240 kA prebaked cell, the anode specification is 1520 × 585 × 535 (mm), and the number of anode carbon block can be calculated as 40. If each group has four pieces, there are ten anode groups.

According to the position of the feed (addition of alumina), prebaked batteries can be classified into two types: side feeding and center feeding. The structure of latter one is shown in Figs. 1.4 and 1.5.

1.3.2 Cathode Device

The cathode device consists of three parts: a steel shell, cathode carbon blocks, and insulating materials. Its structure is shown in Figs. 1.6 and 1.7.

Steel shell: the aluminum reduction cell shell is made of welded steel plates. In order to enhance the strength of the steel shell, reinforcing materials are used around and at the bottom. Figure 1.7 shows the finished shell.

The entire steel shell is mounted on the cement substrate, and electrical insulation material is placed between the shell and the substrate to ensure safe production.

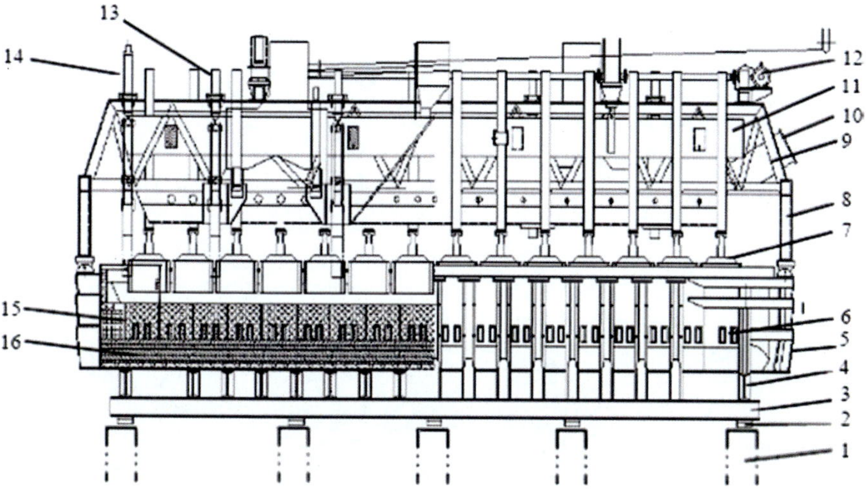

Fig. 1.4 Structural diagram of an aluminum reduction cell with prebaked anodes and amperage of 200 kA. (1) Concrete support; (2) insulation block; (3) L steel; (4) L steel; (5) steel shell; (6) cathode opening; (7) anodic carbon block group; (8) load-bearing support; (9) load-bearing support; (10) anode gas exhaust pipe; (11) anode busbar; (12) anode lifting mechanism; (13) crust breaking device; (14) shell beating device; (15) cathode carbon block group; (16) cathode lining

Fig. 1.5 200 kA prebaked anode aluminum reduction cell in operation

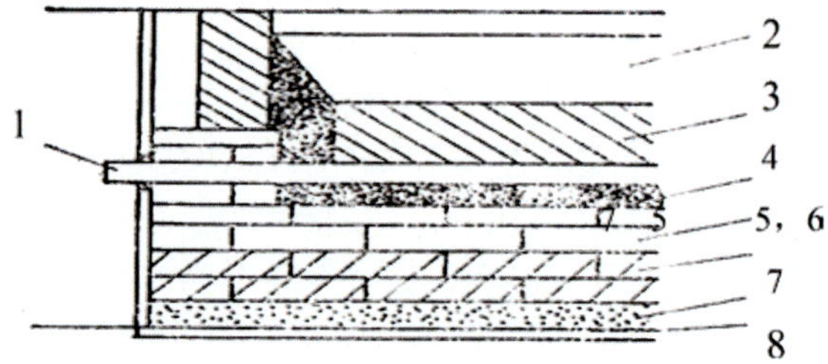

Fig. 1.6 Structure diagram of the cathode device. (1) Cathode rod; (2) electrolytic cell chamber; (3) cathode carbon blocks; (4) carbon cushion; (5, 6) insulation bricks; (7) alumina powder; (8) asbestos board

Fig. 1.7 The steel shell of finished aluminum reduction cell

1.3.3 Conductive Busbar System

The conductive busbar system of an aluminum reduction cell includes anode busbar, cathode busbar, anode risers, and connecting busbar between slots, all of which are large, cast aluminum plates. In addition, there is a flexible anode busbar and a small cathode busbar. The former is used to connect the column to the anode bus, and the latter is used to connect the cathode steel rods and the cathode bus.

The most important part of the conductive busbar system is the configuration of the busbars and the choice of the economic current density of the busbar. The former depends on the requirement of controlling the distribution of the magnetic field of the electrolytic cell, and the latter depends on the optimized result of the electric energy consumption and the investment in the capital construction. There are two types of bus configuration: vertical one and horizontal one. Modern large-scale pre-baked cells are configured horizontally. Figure 1.8 shows examples of two common bus configurations.

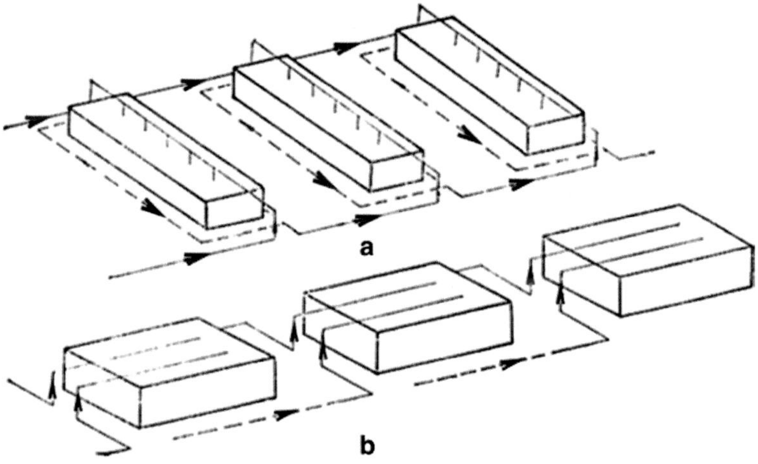

Fig. 1.8 Busbar configuration of aluminum reduction cell. (**a**) Double terminal feed; (**b**) single terminal feed

1.4 Aluminum Reduction Cell Series and Potroom Building

1.4.1 Aluminum Reduction Cell Series

In aluminum production, the series of aluminum electrolytic cells consist of many identical series-connected batteries. The current of each cell is equal, and the number of cells depends on the required production capacity and current intensity. In addition, it is related to the power supply and rectification. The electrolytic aluminum plant can be one series of cells and can also have multiple series. For example, if the current intensity of the series is 80 kA and the current efficiency is 88% in an aluminum plant with an annual output of 30,000 t of aluminum, a series of about 150 electrolytic cells can be formed.

1.4.2 Aluminum Potroom

The factory building is called a potroom. The cathode busbar system and the steel shell are located downstairs in the basement. This structure is always used for modern large prebaked cells. The whole potroom should be bright and well ventilated. There are passage ways in the building, vehicles, and machinery. If there is more than one potroom in the plant, they are located parallel to each other. There are roads between the potroom and the foundry (cast house). There are large aluminum storage areas.

1.5 Carbon Anode Production Process

Aluminum electrolysis requires a large consumption of anode carbon materials, usually 400–450 kg of carbon per ton of aluminum production. Carbon materials used in aluminum electrolysis are different in specifications and types due to their application and requirements, but the production process is similar. The carbon materials for aluminum electrolysis mainly include anode paste, prefabricated anode blocks, sidewall carbon blocks, and bottom carbon blocks. This section focuses on the production process of carbon anode materials, anode paste, and prefabricated anode blocks.

1.5.1 Aggregate—Calcined Petroleum Coke and Coal Tar Pitch

The aggregate of the carbon anode is usually made from a mixture of calcined petroleum coke and coal tar pitch. Delayed petroleum coke (delayed coke) is used in some carbon plants. It is a product obtained by high temperature heating of oil slag and delayed coking process. Pitch is the product of coal bitumen distillation and coking.

The main products for carbon anodes are anode paste and prebaked anode blocks. The main processes of their production include raw material preparation, calcination, crushing and screening, blending and kneading, raw anode block forming, raw block baking, and anode assembly (rodding), as shown in Fig. 1.9.

1.5.2 The Quality Standards of Carbon Anode

1.5.2.1 Anode Paste

The anode paste is used in aluminum electrolysis cells with a self-baking anode. The oxygen released in the electrolysis process reacts with the anode carbon to form CO_2 and CO and causes continuous anode consumption. The quality of the anode paste has influence on the electrolysis process, the energy consumption, and the product quality. At present, aluminum plants have specifications for the ash content of the anode paste and the resistivity, compressive strength, and true porosity of sintered samples. Soderberg cells are classified into horizontal and vertical stud cells (HSS and VSS). In terms of plastics, the requirements of the HSS cells are different from that of the VSS cells. The former requires less fluidity of the anode paste, and the latter is more suitable. Some aluminum factories believe that in addition to the above requirements, the elongation, bulk density, and shrinkage of the anode paste should be considered.

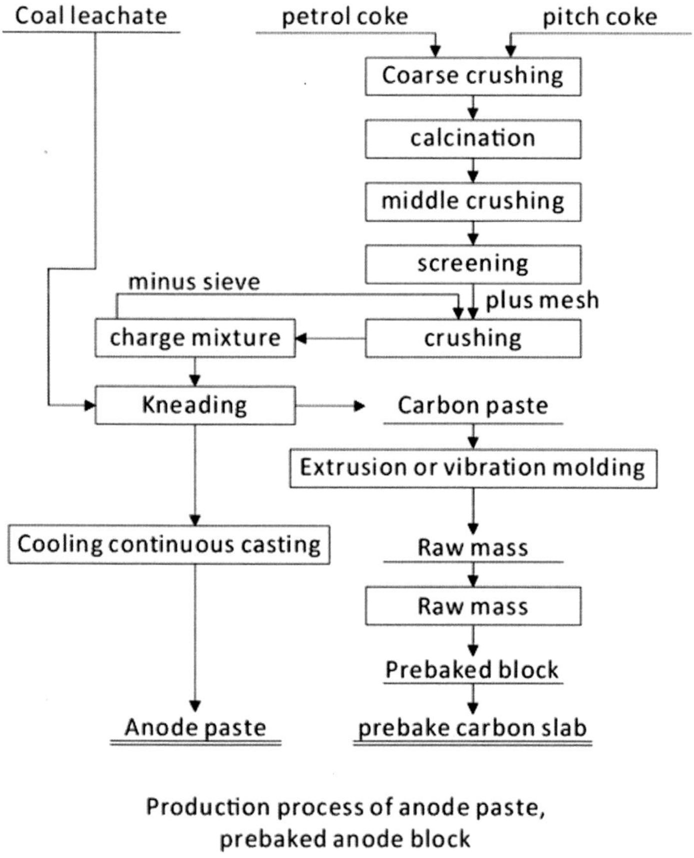

Fig. 1.9 The production process of Soderberg anode paste and prebaked anode blocks

1.5.2.2 Prefabricated Anode Block (Carbon Anode)

For prebaked anodes, quality inspection projects also have requirements, and relevant standards are available. The thermal load and mechanical load of the carbon anode are increased with the large size of modern aluminum reduction cells. Therefore, the flexural strength, thermal conductivity, thermal expansion coefficient, and thermal stress resistance of the carbon anode should be considered.

Cathode carbon blocks and sidewall carbon blocks are both conductive materials and refractories in the cell. The sidewall carbon blocks do not conduct electricity. They are not consumed in the electrolysis process, and their quality has a significant effect on the cell life. To extend its service life, silicon carbide (SiC) is now used as the sidewall material of modern batteries. For cathode blocks, graphite fragments or natural graphite can be added to the ingredients. For example, the addition of semi-graphitized anthracite blocks in Japan can improve the corrosion resistance of electrolyte and aluminum solutions.

1.6 Cryolite and Aluminum Fluoride Production

Cryolite is sodium fluoroaluminate, and its molecular formula is Na_3AlF_6 or $3NaF*AlF_3$. There are two types of cryolite: synthetic cryolite and natural cryolite. It was originally named for its icelike appearance. Cryolite for aluminum electrolysis is now exclusively synthetic cryolite (hereinafter referred to as cryolite), which can be produced by the acid method, the alkali method, the dry method, and the phosphate fertilizer by-product method. Among these methods, the acid method is currently the most widely used.

The process of producing cryolite by the acid process includes acid production, refining of crude acid, synthetic cryolite, filtration and drying of finished product, etc.

The production of 1 t of cryolite by the acid method consumes 0.69–0.71 t hydrofluoric acid, 0.29–0.30 t $Al(OH)_3$, and 0.8–0.9 t Na_2CO_3. The comprehensive energy consumption is about 30–40 GJ, which is equivalent to the energy consumption of production of 1.215 t standard coal.

References

1. Yang, C.Y.: Light Metal Metallurgy. Metallurgical Industry Press, Beijing (2002)
2. Qiu, Z.X.: Prebaked Anode Electrolysis Cell Aluminium Smelting. Metallurgical Industry Press, Beijing (1982)
3. Grjotheim, K., et al.: Aluminium electrolysis. In: The Chemistry of the Hall-Heroult Process. Metallurgical Industry Press., (Chinese translation), Beijing (1982)
4. Grjotheim, K., et al.: Aluminium Smelter Technology – A Pure and Applied Approach. Metallurgical Industry Press. (Chinese translation), Beijing (1985)
5. Qiu, Z.X.: Physical Chemistry of Aluminum Metallurgy. Metallurgical Industry Press, Beijing (1985)
6. http://www.chalco.com.cn
7. http://www.DALiLVCAI.com

Chapter 2
Aluminum Electrolysis Production Process

2.1 Normal Production of Aluminum Electrolysis Cells

The aluminum reduction cell is preheated before starting. Then, it is produced stably under the rated current intensity during the working time of 1 month. Then, all the technical parameters meet the design requirements. Good technical and economic results are obtained. This mainly refers to current efficiency, electric energy consumption, the quality of the primary aluminum produced, etc.

Another important feature of the cell is the shape of it. In the electrolysis cell, there is a ring of crust (solidified electrolyte), which is known as the side ledge. The side ledge can effectively prevent the loss of electric energy and heat energy passing through the sidewall and protects the inner lining of the sidewall. During the electrolysis process, efforts should be made to maintain the regularity of the side ledge. The side ledge is shown in Fig. 2.1.

In addition, the bottom of the cell should be clean in normal production, that is, no or only a small amount of alumina precipitation. This is called sludge. The maintenance of normal production depends on reasonable technical conditions and a corresponding good operating system, as well as meticulous degree of operation. Otherwise, the production will not be maintained normally, resulting in a decline in various technical and economic indicators.

2.2 Technical Conditions for Normal Production of Aluminum Electrolysis

The normal technical operating parameters are based on aluminum electrolysis theory and specific conditions.

The technical conditions for the normal production of aluminum electrolysis mainly include amperage, cell voltage, electrolyte temperature, interpolar distance,

W. Xianxi, *Inert Anodes for Aluminum Electrolysis*, The Minerals, Metals & Materials Series, https://doi.org/10.1007/978-3-030-28913-3_2

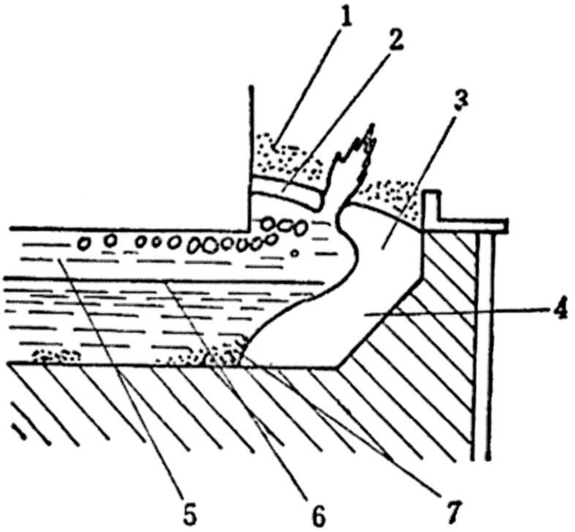

Fig. 2.1 The side ledge of an aluminum reduction cell during normal production. (1) Al_2O_3 powder on top of the anode cover; (2) the anode cover itself; (3) side ledge; (4) extending side ledge in the metal phase; (5) electrolyte; (6) molten aluminum; (7) bottom sedimentation (undissolved alumina, also called sludge)

electrolyte composition, the height of the electrolyte layer and the height of the molten aluminum layer, cathode voltage drop, and anode effect frequency and duration. The above technical conditions, or technical operating parameters, are all interrelated. In a certain period of time, the relative stability should be maintained as well as possible. When changes are needed for some reasons, the remaining parameters must be adjusted accordingly.

2.3 Conventional Operation of Aluminum Electrolysis

As the main contents of conventional operation of aluminum electrolysis, both the self-baked Soderberg battery and the modern prebaked battery include three main parts: alumina feed, aluminum removal (tapping), and anode replacement. The specific content varies according to the battery type.

2.3.1 Feeding of the Raw Material

The feeding operation of the aluminum reduction cell is to add alumina to the electrolytic cell on a regular and quantitative basis. The modern cells now adopt nearly continuous or so-called point feeding of alumina, and there is a specified time gap between each addition, which may be only as low as about 20 s.

The purpose is to keep the concentration of Al_2O_3 in the electrolyte constant and stable. The concentration of Al_2O_3 in the electrolyte used to be between 4% and 5%. However, in recent years, the concentration of Al_2O_3 is generally only between 2% and 3% or even lower. Because of the low alumina concentration, the current efficiency can be up to 95%. In addition, Al_2O_3 can be quickly dissolved in the electrolyte.

Most of the alumina feeding of the cell is controlled by a computer. The feeder positions are in the center slot of the cell. The feeding procedure of aluminum electrolysis is to open the alumina cover by a crust breaker, and then a batch of alumina is added to the newly opened feeding hole in the crust. In addition, a feeding operation is sometimes accompanied by the adjustment of the electrolyte composition, for example, by adding additives for adjusting the molar ratio of the electrolyte.

2.3.2 *Aluminum Tapping*

The aluminum produced during the electrolysis process is periodically and quantitatively removed from the electrolytic cell. The time between two tapings is usually 1–2 days, and large cells can be tapped once a day. The tapping is carried out with a vacuum ladle. The extracted aluminum is shipped to the foundry (cast house) for treatment. Before the aluminum ingot is cast, the molten aluminum needs a series of treatments, such as flux purification, quality blending, slag clarification, etc. These processes are usually carried out in a certain order in the mixing furnace. Then, the aluminum is cast into aluminum billets or commercial aluminum ingots of various shapes and sizes.

2.3.3 *Anode Changing*

The anode can be called the heart of the electrolytic cell, and its management is very important. The anode operation depends on the cell type. There are three steps of prebaked anode operations:

1. The prebaked anodes are replaced regularly in a certain order to keep the new and old anodes (not yet replaced) evenly sharing the current and ensuring that the anode beam is not inclined.
2. The anode must be covered with alumina after replacement (Fig. 2.2).
3. The anode beam has to be raised because the position of the beam decreases with the gradual consumption of the anode blocks.

The above are the three main steps of operations of prebaked aluminum electrolysis cells, and the operation quality is the precondition to ensure the technical quality of aluminum electrolysis. In addition, the process of aluminum electrolysis includes the work of extinguishing anode effects, the treatment of poorly operating cells, etc.

2.3.4 *Computer Control of Aluminum Reduction Cells*

The application of computers in the process of aluminum electrolysis is a remarkable achievement in the development of aluminum production. It increases the possibility of production under the optimal conditions, thereby greatly improving

Fig. 2.2 Replacement of the prebaked anode in the cell and covering of the anode with alumina on all sides

production parameters. For example, the current efficiency remained between 87% and 90% in the 1950s and 1960s, exceeded 90% in the late 1970s, and increased to 94–95% in recent years. K. Grjotheim believed that when conditions are optimal, the current efficiency may reach 98%. Obviously, the realization of this optimal condition depends on the use of computers.

Since the late 1960s, the aluminum electrolysis industry has adopted computer control technology. Since the 1970s, all countries have realized the advantage of computer control of electrolytic aluminum production. The continuous improvement of the function of computer control not only liberates the operator from the heavy physical labor in the working environment of high temperature, strong magnetic fields, and high dust but also realizes accurate, timely, and stable cell control. It is also possible to use modern (180–600 kA) prebaked anode cells for electrolysis under the conditions of low temperature, low molar ratio, and low concentration of Al_2O_3. This can improve current efficiency and reduce energy consumption. A current efficiency of 94% and direct energy consumption of 13,000–13,300 kW h/t Al are obtained under the management of advanced computer systems. With the improvement of the automatic operational level of aluminum reduction cells, it is possible to design a closed electrolytic cell to a high degree, thereby pushing the aluminum electrolysis production to the direction of low pollution. With the use of computer systems, the traditional management model of aluminum electrolysis is rapidly becoming digital, standardized, and intelligent. The computer control and management system has become an indispensable automation technology in the modern aluminum electrolysis production process and has become one of the important symbols of the development level of contemporary aluminum smelting technology.

For large cells and their series, it is very convenient to use computers to control the production condition and operation. Small cells are more stable. Thus, abnormal situations are difficult to appear, such as sick cells, etc. This provides good conditions for computer control. Highly mechanized operations can be performed under the control of a computer, such as shell breaking and alumina feeding. However, due to the high temperature of the electrolyte and its strong corrosion ability, many parameters cannot be continuously measured (such as electrolyte temperature, alumina concentration, etc.) and cannot be directly controlled. Thus, most processes are mainly controlled by controlling the cell voltage.

2.4 The Gas Purification of Aluminum Plant Anode

2.4.1 Pollutants in the Anode Gas of Aluminum Plants

The anode gas contains HF, SiF_4, CF_4, CO_2, CO, bitumen volatiles, and various types of fluoride dust, which cause damage to trees, forage, and livestock in the surrounding environment of the aluminum plant. The potroom operators are also at risk. Therefore, the gas treatment of aluminum plant now attracted more and more attention. Many countries have strict legislation on fluoride emissions in the anode gases from aluminum plants. The aluminum plant has been forced to clean the anode gases to meet requirements of increasingly stringent environmental protection. In the aluminum electrolysis process, there are two kinds of pollutants in the anode gas: gaseous pollutants and solid particle pollutants.

1. Pollutants as gaseous substances

   The gaseous pollutants include the CO_2, CO, the gas produced during the anodic process, the anode effect gas CF_4, volatilization of asphalt smoke produced by self-baking Soderberg anodes, hydrogen fluoride gas produced by hydrolysis of fluoride salts such as aluminum fluoride, impurities, SiF_4 in raw materials, etc. The main gas pollutants are HF and other fluorinated gases.
2. Pollutants as solid particulates

   Particulate pollutants are mainly produced by volatilization of raw materials, including solid dust such as Al_2O_3, C, Na_3AlF_6, and $Na_5Al_3F_{14}$. The main pollutants are cryolite and HF adsorbed on alumina dust.

The ratio of gaseous pollutants to solid pollutants depends on the cell type. About 60–90% of the pollutants from Soderberg cells are gaseous, and about 50% of the pollutants from prebaked anode cells are gaseous.

2.4.2 Method of Anode Gas Purification in Aluminum Plants

2.4.2.1 Wet Purification

Wet purification is the purification of the anode gas with water or alkali solution as absorbent of HF and dust:

$$6NaF + 4NaHCO_3 + NaAlO_2 \rightarrow Na_3AlF_6 \downarrow + 4Na_2CO_3 + 2H_2O$$

The fluoride-containing gas is usually washed with 5% sodium carbonate solution. When the NaF content in the solution reaches 25–30 g/L, $2Na_2CO_3 + 2HF–2NaF + 2NaHCO_3$ is used to synthesize cryolite.

Cryolite is separated, dried, and returned to the cells for use. The filtrate is returned to the absorption tower to absorb the fluoride gas, and the material of whole process can be recycled.

2.4.2.2 Dry Purification

Dry purification uses alumina as absorbent to adsorb HF in the gas and retain dust. HF adsorbed on alumina is the raw material of electrolysis. This is called secondary alumina. According to reports, the adsorption of HF gas by alumina is mainly chemical adsorption. It is reported that the adsorption of HF gas by alumina is mainly chemisorption. During the adsorption process, monolayers are formed on the surface of the alumina, which adsorbs 2 mol of HF per mole of alumina. The surface compounds are converted to AlF_3 molecules at temperatures above 300 °C:

$$Al_2O_3 + 6HF = 2AlF_3 + 3H_2O$$

The rate of this process is very high; it only takes 0.25–1.5 s to complete. The adsorption efficiency can reach 98–99%. The specific surface area and the surface activity of alumina used in the aluminum industry are different due to different calcination temperatures, which make the adsorption properties of HF different. The specific surface area of alumina for dry purification should be greater than 35 m^2/g, and the content of α-Al_2O_3 should not exceed 25–35%. The specific surface area and α-Al_2O_3 content of sandy alumina are based on this requirement. Dry purification is more suitable for anode gas purification from prebaked anode cells. Figure 2.3 shows a dry gas purification and exhaust device for prebaked anode cells.

2.5 Economic Analysis of Primary Aluminum Production

2.5.1 Cost Analysis of Primary Aluminum Production

The cost of primary aluminum produced by electrolysis can be roughly divided into three parts, namely, raw material cost, electrical energy cost, and production management fee (or labor management fee).

1. The raw materials used in the production of aluminum mainly include alumina, aluminum fluoride, additives, and carbon materials. Their costs account for about 50% of the total cost, and the alumina alone accounts for 40%.
2. The electricity consumption cost of aluminum electrolysis is very high. The electrical energy consumption of aluminum per ton is 13,500–14,000 kW h. If the rectifier power loss, power consumption, and remelting power consumption are taken into account, the electrical energy consumption per ton of aluminum is greater.

Fig. 2.3 Anode gas dry purification and exhaust device of prebaked anode cells in an aluminum plant

When the DC electrical energy consumption rate is 14,000 kW h/t Al, the above-mentioned electrical energy consumption rate is about 15,340 kW h/t Al. According to the calculations of the British Aluminum Plant, in production of 1 t of primary aluminum from bauxite mining to aluminum ingot casting, the total electrical energy consumption rate is about 22,800 kW h/t Al.

3. Due to the different production levels and wages in various countries and regions, the impact on the production cost of raw aluminum is also different. Because raw materials and electricity consumption constitute the main cost of raw aluminum production, this book mainly considers the influence of the above raw materials and electricity on the production cost of raw aluminum.

2.5.2 *Ways to Reduce Costs*

2.5.2.1 Reducing the Loss of Raw Materials

The consumption of alumina per ton in the current aluminum electrolysis plants is 1930–1940 kg, and the theoretical consumption is about 1890 kg. The excess consumption is mainly caused by transport losses, losses during feeding, and other mechanical losses. With the development of mechanization and automation, the loss of alumina can be reduced. For the carbon material, the theoretical consumption of aluminum per ton is 393 kg (the content of carbon dioxide in the anode gas is 70%).

However, the actual consumption of aluminum per ton is as great as 530 kg, and the utilization ratio is only 74% (China). The net consumption of carbon anode of aluminum per ton in western countries is below 410 kg, which is close to the theoretical value. If low-temperature electrolysis is used, the carbon consumption will be reduced further. The use of inert cathode and inert anode cells is also one of the possible ways to reduce the cost.

In theory, cryolite is not consumed in the electrolysis process. However, the volatilization of fluorides (mainly AlF_3) at high temperature, interaction with alkali and alkaline-earth metal oxides, and hydrolysis result in their loss. The main way to reduce its volatilization is to lower the electrolysis temperature.

2.5.2.2 Strengthening Production Management and Extending the Effective Production Time of the Electrolytic Cells

Electrolytic cells are disconnected mainly due to damaged lining and cell overhaul. The cost of cells for overhaul is as high as one-third of the cost of the newly built cells.

Therefore, prolonging the effective production time of the cell can reduce the production cost of the electrolytic cells.

1. Strengthening the production management, perfecting the operation technical conditions, and strictly carrying out the operation conditions, ensuring normal operation, or fewer sick cells, are important measures for prolonging the life of the electrolytic cell.
2. Using graphitized and semi-graphitized cathode blocks, the cell life can be increased from 2000 to 2400 days.
3. The construction of new aluminum plants and the renovation of old ones are done according to the principle of large scale and high efficiency. In this process, low molar cryolite ratio (below 2.2), low alumina concentration (Al_2O_3 of 1.5–3%), and low temperature (below 940 °C) are used together with optimum design of electrolytic cells by mathematical models (including magnetic field, thermal field, and hydrodynamic physical field). The use of computers to control the production process can improve the current efficiency and reduce the production cost. The calculation shows that the investment cost and production cost of large cells decrease with the increase of aluminum plant size. If the annual output is increased from 100,000 to 125,000 t, the unit investment will be reduced by 10%, and the cost will be reduced by 5%.

References

1. Yao, H.B.: Practice of Aluminum Electrolytic Production. Metallurgical Industry Press, Beijing (1984)
2. Yang, C.Y.: Light Metal Metallurgy. Metallurgical Industry Press, Beijing (2002)
3. Qiu, Z.X.: Aluminium Electrolysis. Metallurgical Industry Press, Beijing (1982)
4. Grjotheim, K., et al.: Aluminium electrolysis. In: The Chemistry of the Hall-Heroult Process. Metallurgical Industry Press., (Chinese translation), Beijing (1982)
5. http://www.lvye.biz/
6. http://www.chalco.com.cn

Part II
Review of the Study on the Inert Anode of Aluminum Electrolysis

Chapter 3
Aluminum Electrolytic Inert Anode

3.1 Disadvantages of Current Hall-Heroult Aluminum Electrolysis Process

In the traditional Hall-Heroult molten salt aluminum reduction cell, Na_3AlF_6-based fluorine salt melt is used as solvent to dissolve Al_2O_3 in fluoride salt melt to form oxygen-containing complex ion and aluminum complex ion. Due to the strong corrosion at high temperature (usually 940–960 °C) with the exception of precious metals, carbon materials, and very few ceramic materials, most materials have high solubility. Since the invention of Hall-Heroult molten salt aluminum electrolysis process, carbon material has been used as anode material and cathode material. When a direct current is connected between the carbon anode and the carbon cathode, the Al-containing complex ion discharges on the surface of the cathode (actually liquid metal), and the metal aluminum precipitates. The oxygen containing complex ions discharges on the surface of carbon anode immersed in electrolyte melt and combines with carbon anode to form CO_2 precipitation. The electrolysis process can be expressed simply as:

$$Al_2O_3 + \frac{3}{2}C = 2Al + \frac{3}{2}CO_2 \uparrow \tag{3.1}$$

3.1.1 *Carbon Anode Consumption*

In Formula (3.1), the carbon anode is expendable in the electrolysis process. Therefore, the carbon anode must be replaced periodically. This brings many problems.

1. Consumption of high-quality carbon materials. If the current efficiency is 100, the carbon content of the anode is 100, calculated by Formula (3.1). The

W. Xianxi, *Inert Anodes for Aluminum Electrolysis*, The Minerals, Metals & Materials Series, https://doi.org/10.1007/978-3-030-28913-3_3

theoretical carbon anode consumption of ton of aluminum is 333 kg. However, due to the secondary reaction of Al (current efficiency less than 100%), air oxidation of carbon anode, and the removal of carbon slag, the actual net consumption of carbon anode exceeds 400 kg.

2. Environmental pollution. The equivalent CO_2 emissions per ton of aluminum from the current Hall-Heroult aluminum electrolysis process are shown in Table 3.1. Among them, the aluminum electrolysis process produces a large amount of greenhouse effect gas (GHG) or harmful gas, which mainly includes three parts:

   (a) Carbon-containing compounds (CO_2 and a small amount of CO) produced during electrolytic reaction
   (b) Fluorocarbon CF_n released with the appearance of anodic effect
   (c) HF produced by raw material containing H_2O reacting with fluoride electrolyte

The equivalent carbon dioxide emitted by the electrolytic reaction comes from three main sources: (a) the anode reaction produces CO_2 1.2 kg/kg (Al); (b) the extra oxidation of anode to produce CO_2 0.3 kg/kg (Al); and (c) the electrical energy consumption per ton of primary aluminum electrolysis (15,000 kw h). According to the type of energy used, the emission of CO_2 during power generation is 13.5 kg/kg (Al). Therefore, the equivalent CO_2 per ton of aluminum production is 6.3 t.

When anodic effect occurs, the CF_n emissions are mainly CF_4 and C_2F_6, and the GWP (global warming potential, which is used to characterize the relative greenhouse effect of various gases relative to CO_2) of these two greenhouse gases reaches 6500 and 9200, respectively. The equivalent greenhouse effect of anodic effect gas (the average value of producing CO_2 2.0 kg/kg (Al)) is mainly determined by the anode effect coefficient and effect time, which mainly depends on the electrolytic cell structure, especially the feeding mode and its control system.

Table 3.1 Equivalent CO_2 emissions of per ton of aluminum during Hall-Heroult aluminum electrolysis process [1]

Production process	Hydropower or nuclear power	Natural gas fire point	Coal thermal power generation	World average
Bauxite and Al_2O_3 production	2.0	2.0	2.0	2.0
Carbon anode production	0.2	0.2	0.2	0.2
Electrolytic process	1.5	1.5	1.5	1.5
Anode effect	2.0	2.0	2.0	2.0
Generate electricity process	0	6.0	13.5	4.8
Total release	5.7	11.7	19.2	10.5

According to the consumption of carbon anodes per ton of aluminum, CO_2 is produced during the production of carbon anodes. The corresponding CO_2 emissions from carbon anode production can be calculated as 0.2 t. In addition, a large amount of asphalt fumes, mainly polycyclic aromatic carbohydrates, are produced in the process of carbon anode production, which results in the environmental pollution.

3. Affecting the stability of the normal operation of the electrolytic cell. First, the current distribution and thermal balance of the electrolytic cell are disturbed by the frequent replacement of the anode. The maintenance and replacement of the anode require more working hours and labors, which increases the production cost. Second, the carbon slag appears in the electrolyte due to the uneven oxidation and caving of the carbon anode.

3.1.2 *Poor Wettability of Carbon Cathode and Liquid Aluminum*

Carbon material has been used as the cathode material in aluminum reduction cells. Due to the poor wettability of the metal aluminum liquid and the carbon cathode material surface, the electrolytic cell must maintain a certain level of aluminum liquid to prevent the carbon cathode surface from being exposed to the electrolyte (to avoid the continuous formation and dissolution of Al_4C_3). The liquid aluminum moves under the action of electromagnetic force, causing the deformation and fluctuation of the interface between liquid aluminum and electrolyte. The lower the height of liquid aluminum is, the more violent the motion of liquid aluminum is. Therefore, the height of aluminum liquid in current aluminum reduction cells must be kept above 15 cm. To prevent the movement and interface deformation of molten aluminum from affecting the current efficiency, the electrolytic cell has to maintain a higher pole distance (4 cm). It is an important reason for the current aluminum reduction cell to maintain a high cell voltage and high energy consumption. According to the calculation statistics, the voltage drop between the two poles of aluminum reduction cell is 1.3–2.0 V. Comparing the electrochemical theoretical decomposition voltage of aluminum electrolysis process with that of 1.2 V, a large part of the energy of the current aluminum electrolysis process is consumed between the two poles in a Joule heat manner. If the polar distance can be reduced, both the energy consumption of aluminum electrolysis and the production cost of primary aluminum can be reduced.

In addition, metal aluminum and carbon cathode can react to form Al_4C_3 at electrolysis temperature. When the liquid aluminum does not cover the cathode well, Al_4C_3 will directly contact with the electrolyte and dissolve into it. This promotes the formation of Al_4C_3 and the corrosion of the cathode, which seriously affects the life of the electrolytic cell.

3.1.3 Other Problems with Carbon-Lined Materials

In the process of aluminum electrolysis, both metal aluminum and sodium metal are precipitated on the surface of the cathode. When the modern prebaked aluminum electrolysis cell is started, the molten cryolite electrolyte is first poured into the electrolyzer, and the precipitation of sodium is especially rapid. In addition, the replacement reaction of aluminum with NaF can also produce Na. Sodium permeates into the cathode carbon materials to form intercalation compounds, resulting in cathodic volume expansion or even cracking. This has become a major cause of cell breakage, which increases the investment in aluminum electrolysis plants and the production cost of primary aluminum.

A large number of waste inner linings are produced after the aluminum reduction cell is damaged. According to the current life estimation of the electrolytic cell, 1 t of aluminum metal production will produce 30–50 kg of waste lining. In addition to about 30% of carbonaceous materials, the waste liner contains cryolite, sodium fluoride, nepheline, sodium aluminum oxide, a small amount of α-alumina, aluminum carbide, aluminum nitride, aluminum ferroalloy, and trace cyanide. The waste lining of aluminum reduction cell is a polluting solid waste, among which cyanide is a highly toxic substance and sodium fluoride has a strong corrosion. When waste lining meets water (such as rain water, surface water, groundwater), sodium fluoride and cyanide will dissolve in water, causing F^- and CN^- to mix into rivers and to seep into the ground. This will pollute water source and cause long-term serious pollution to the surrounding. Therefore, researchers have been carrying out research to solve or alleviate the problems caused by it. Most of them use high-temperature incineration waste lining to remove toxic chemicals, recover valuable fluorides (such as AlF_3), and make the residual substances chemically inert.

In addition, the traditional aluminum reduction cell adopts carbon material as the inner lining of the side wall. To reduce the side oxidation and conduction, the side heat dissipation is required to form the side shell, which leads to the energy consumption.

3.1.4 Horizontal Structure of Hall-Heroult Electrolyzer

Carbon anode and carbon lining of surface level are used as cathodes in current Hall-Heroult electrolytic cells. The aluminum precipitated by electrolysis accumulates above the carbon cathode at the bottom of the cell to form an aluminum dissolving cell as the actual cathode. The anode is fixed with a clamp, and its guide rod is suspended from the anode beam busbar in the upper part of the groove. The lower end of the carbon anode is immersed in the electrolyte in the cell and is close to the surface of the molten aluminum at the bottom of the cell. The square steel is embedded in the cathode carbon block, one end of which extends out of the groove and is connected with the external cathode busbar. The current flows from the busbar outside the trough to the soft strip bus and then to the crossbar. After that, the current flows through the anode to the electrolyte and liquid aluminum and then from the cathode through the cathode steel bar to the cathode bus connected to the column bus of the next slot. This is a complete current channel. The size and electrolysis process of the existing Hall-Heroult aluminum reduction cell are different. There is a problem of low power efficiency, which is generally about 45%. In addition to the energy needed to reduce alumina to aluminum in theory, the other electrical energy in the actual electrolytic production is dissipated in the form of thermal energy. The main reason for the difference between the theoretical energy consumption and the actual energy consumption is that the current Hall-Heroult aluminum reduction cell adopts the horizontal structure mentioned above. The high pole distance operation results in low production capacity and high cell voltage.

There are two ways to save energy and reduce consumption in aluminum reduction cells. One is to improve current efficiency, and the other is to reduce cell voltage and pole distance. However, the current efficiency of large prebaked aluminum reduction cells has reached 95% or more, and reducing energy consumption by the improvement of current efficiency will not be too strong. In addition, the current pole distance of the prebaked cell is usually greater than 4 cm. This makes the pressure drop between the poles reach 1.3–2.0 V. It provides a lot of space for reducing energy consumption by reducing the distance between the poles. However, for common prebaked cells, the reduction of the pole distance affects the thermal balance of the electrolytic cell. In addition, the polar distance cannot be reduced too much, even within the allowable range of the thermal balance. This is mainly because the reduction of the pole distance leads to electrolytic instability. This causes the molten aluminum to fluctuate, reducing the current efficiency, and the gain is not worth the loss. In order to reduce the pole distance and the energy consumption of aluminum reduction cells, it is necessary to improve the cell and adopt new materials, inert cathodes, and the new type of electrolytic cell.

3.2 Study on Inert Anode

3.2.1 Advantages of Inert Anodes

Aluminum electrolytic inert anode refers to the anode that is not consumed or consumed very slowly. When inert anodes are used, oxygen is released from the anode. The equation of the aluminum electrolysis process is:

$$Al_2O_3 = 2Al + \frac{3}{2}O_2 \uparrow \tag{3.2}$$

In Formula (3.2), the inert anode is not consumed during the electrolysis process, and disadvantages caused by the consumable carbon anode are absent. Compared with carbon anode, the inert anode has many advantages, such as environmental protection, energy saving, simplification of operation, and cost reduction, especially the potential to reduce pollution and the cost of primary aluminum production, as shown in Table 3.2 [2].

However, the inert anode has disadvantages, that is, the reversible decomposition voltage of reactive Formula (3.2) is high. The reversible decomposition voltage of reactive Formula (3.2) at 1250 K is 2.21 V. The reversible decomposition voltage of reactive Formula (3.1) is only 1.18 V at the same temperature. In other words, the reversible decomposition voltage is reduced by 1.03 V with the use of carbon anode. However, this reduction requires the consumption of carbon materials. In addition, the high decomposition voltage of Al_2O_3 on the inert anode can be compensated by other parts in Table 3.3. It can still achieve the purpose of energy saving. J. Noel pointed out that [3] if we do not change the anode distance in the case of inert anodes, 5 cents can be saved; if we change the distance between the anode and the cathode, 23 cents can be saved; and if we use a wettable cathode (TiB_2 inert cathode) and change the distance between the electrodes, then the maximum energy saving can be up to 32%. The comparison of voltage drop and energy consumption between a new type of electrolytic cell with inert anode and current Hall-Heroult cell are shown in Table 3.3.

When the inert anode is used in the aluminum reduction cell, the discharge of CO_2, CO, and CF_n is absent in the aluminum electrolysis process, and the O_2 is absent from anode emission (which can be used as a by-product). Using inert anodes, the equivalent CO_2 emission of aluminum produced by global aluminum electrolysis will be reduced from 10.5 t to 7.1 t, a reduction of nearly 32%.

Table 3.2 Advantages of inert anode in aluminum electrolysis [2]

Environmental protection	Cost/capacity	Energy/ consumption	Process/ control	Safety/health
1. Reducing, eliminating CO_2 emissions 2. Elimination of PFC emissions 3. Elimination of bituminous flue gas emissions 4. Elimination of bituminous flue gas emissions 5. Eliminating dust emission from coke and paste during anode roasting 6. Reducing the generation of used lining 7. Reducing HF emissions is more flexible	1. Reducing the cost of anode manufacturing 2. Improving the quality of aluminum production 3. Increase the utilization ratio of electrolytic cell space 4. Increase the unit volume capacity of electrolytic cell 5. Reduce operational manpower 6. Slot structure design	1. Improving the thermal efficiency of electrolytic cell 2. Saving energy in carbon anode 3. More energy saving in anode production 4. The electrode spacing and energy can be greatly reduced by matching with wettable cathode	1. Elimination of carbon anode production plant 2. Reducing the frequency of anode replacement 3. Anode bottom is flatter 4. Better control of pole distance	1. Reduce anode replacement operation 2. Electrolytic cell is more closed 3. Improving the working environment of electrolytic workshop

Table 3.3 Voltage drop of aluminum electrolytic cell using different electrodes based on the current efficiency of 91% [3]

Voltage and energy	Hall-Heroult cell	Inert anode electrolyzer		
Consumption	Polar distance 4.45 cm	Polar distance 4.45 cm	Polar distance 1.91 cm	Polar distance 0.64 cm[a]
External pressure drop/V	0.16	0.16	0.16	0.16
Anode connection voltage drop/V	0.16	0.16	0.16	0.16
Anode voltage drop/V	0.16	0.16	0.16	0.16
Electrolyte pressure drop/V	1.76	1.76	0.75	0.26
Decomposition voltage/V	1.20	2.20	2.20	2.20
Polarizing voltage/V	0.60	0.15	0.15	0.15
Cathode drop/V	0.60	0.60	0.60	0.60
Total trough voltage/V	4.64	5.19	4.18	3.68
Direct current consumption kW h/kg	15.2	16.96	13.66	12.0
Total energy saving %	–	5.4[b]	23[b]	32[b]

[a]Matched TiB_2 cathode
[b]Including energy savings in anode production and carbon-free anode consumption

Considering the reduction in energy consumption of tons of aluminum, the equivalent CO_2 emissions will be further reduced.

3.2.2 *Performance Requirements and Research Survey of Inert Anode*

3.2.2.1 Definition of Inert Anode

Inert anodes refer to those anodes that are not or micro consumed in the cryolite-alumina molten salt electrolysis. In recent years, many works have reported the research data of the inert anode, aiming to replace the carbon anode (active anode).

When using carbon anode electrolysis, the reaction formula of aluminum electrolysis is as follows:

$$Al_2O_3(s)+\frac{3}{2}C(s)\rightarrow 2Al(l)+\frac{3}{2}CO_2(g) \tag{3.3}$$

At the temperature of 1300 K, the decomposition voltage of Al_2O_3 is 1.18 V, the theoretical carbon consumption is 0.333 kg/kg (Al) (calculated as 100% current efficiency), and the actual carbon consumption is 0.45–0.55 kg/kg (Al). Due to the requirement of aluminum purity, the theoretical energy consumption of oil coking anode aggregate is 6.5%, 32 kw h/kg (Al) (based on current efficiency of 100%).

The actual energy consumption of the new large cell is 13.5%, 5 kw h/kg (Al). If the inert anode is used, the aluminum electrolysis reaction formula is:

$$Al_2O_3(s) \rightarrow 2Al(1) + \frac{3}{2}O_2(g) \tag{3.4}$$

At 1300 K, the decomposition voltage of Al_2O_3 is 2.2 V. The anode does not participate in the reaction and does not provide energy. Thus, theoretical energy consumption is as high as 9.24 kw h/kg (Al). In theory, the difference in Al_2O_3 decomposition voltage between inert anode and active anode is 1 V, and the difference in power consumption between them is 3 kw h/kg (Al). The reason is as follows. (a) When the carbon anode is applied, the bubble on the anode is larger, and the anode overvoltage reaches 0.4–0.6 V. When the inert anode is applied, the bubble on the anode is smaller, and the anode overvoltage is only 0.2 V. This will bring back 0.3 v. (b) The greatest component producing the cell voltage is the distance between cathode and anode, which is 4–5 cm. The voltage drop reaches 1.6–1.8 V. If the inert anode is combined with the inert cathode to form a new type of electrolysis cell, the inter-pole ancestor can be reduced from 4–5 cm to 3 cm, which is equivalent to reducing the cell voltage by 0.5 V. (c) Save expensive oil coke for anode manufacturing. (d) It is unnecessary to replace anodes frequently, thus reducing operating costs; (e) The cost of aluminum production using inert anode is approximately the same as that using prebaked carbon anode, or slightly lower. However, the investment cost and anode consumption are greatly reduced.

3.2.2.2 The Criteria for Inert Anode Materials

The main criteria are as follows:

1. Selecting the material with low solubility in cryolite-alumina melt to ensure the quality of aluminum
2. Choosing the material with weak reaction ability to aluminum liquid or anode gas O_2
3. Selecting cheap and easily available materials
4. Selecting materials that can be machined into large inert anodes and easily connected with conductors
5. Selecting materials that are well wetted by cryolite melt and are not easy to produce anode effect

Precious metal materials (such as platinum and gold) can be used as inert anodes in laboratories. However, they are not suitable for industrial aluminum production. Nickel and copper with good conductivity cannot be used as anode material alone, and they can only be mixed with other materials to form ceramic anode. Some metal oxides have relatively low solubility in cryolite and alumina melts and can form composite ceramic materials for testing to achieve good electrical conductivity and appropriate structural stability. It has enough durability to ensure the quality of the product aluminum, but it is difficult to find a completely suitable material. A variety

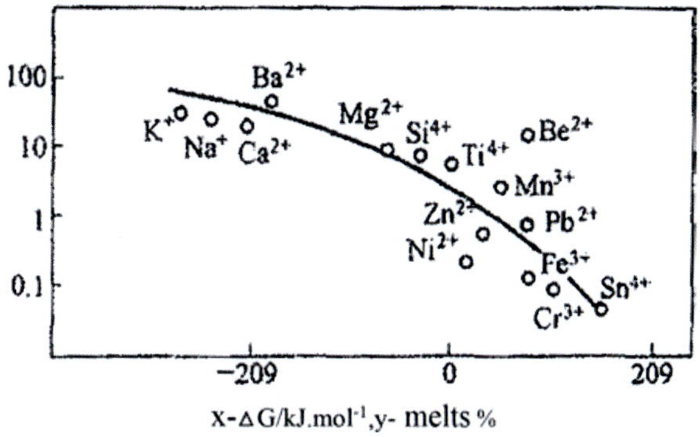

Fig. 3.1 Solubility of metal oxides in cryolite solution [4] x $\triangle G$/kJ/mol, y solubility (%)

of candidate metal oxides are listed in Fig. 3.1, including tin oxide, nickel oxide, iron oxide, antimony oxide, etc., which are characterized by low solubility in cryolite melts and good electrical conductivity at high temperatures.

The general formula for the dissolution of metal oxides in cryolite melts is:

$$\frac{1}{y}M_xO_y + \frac{2}{3}Na_3AlF_6 = \frac{x}{y}MF_{\frac{2y}{x}} + \frac{1}{3}Al_2O_3 + 2NaF \tag{3.5}$$

The Swiss aluminum company proposed to use tin oxide (SnO_2)-based anodes doped with Fe_2O_3, Sb_2O_3, or CuO. The American Diamond Institute proposed to use yttrium oxide (Y_2O_3)-based anodes with additives including precious metal oxides. Sumitomo Corporation of Japan proposed to use the spinel-type, rutile-type inert anode.

Without suitable anode materials, the inert anode can be corroded in the process of aluminum electrolysis. Therefore, the inert anode has not yet been applied in industry. Alcoa suggested that the ideal annual thickness consumption of inert anodes should not exceed 3–5 mm, which will not affect the grade of aluminum in cathode products. The ideal goal is to find a conductive anode material that can resist oxidation and is not corroded by cryolite and aluminum.

3.2.3 *Certain Candidate Materials for Manufacturing Inert Anodes*

In 2002, R. P. Pawlek et al. [5] summarized the literature on inert anode in recent years. In addition, they preliminarily considered that the following kinds of materials are candidates for making inert anode:

3.2.3.1 Oxide Materials

1. $NiFe_2O_4 \rightarrow CoFe_2O_4$
2. NiO - Li_2O

3. SnO_2 $: 96\%SnO_2 + 2\%Sb_2O_3 + 2\%Cu,$
   $: 96\%SnO_2 + 2\%Sb_2E_3 + 2\%AgO$
4. $ZnFe_2O_4$

3.2.3.2 Ceramic Materials

1. Ni-Al-Fe-Cu-X alloy
2. Ni + 6Al + 11Fe+10Cu + 3Zn
3. Ni-Fe alloy
4. Oxygen ion conducting film

Rapp has developed a new non-consumable anode, which is a thin dense oxygen ion conductive material. It can be electrochemically oxidized into natural gas and can reduce CO_2 emission by 50%.

Sadoway et al. of the Massachusetts Institute of Technology (MIT) have studied the application of alloy materials in the preparation of inert anodes, because metal alloys are easy to cast and have good electrical conductivity. Many authors believe that the inert anode should be used in combination with low-temperature aluminum electrolytes. The electrolytic temperature is 730–910 °C, and the electrolyte should be saturated with Al_2O_3.

Duruz and De Norar [6] used Ni-30% Fe Alloy as anode, which is oxidized in air at 1100 °C for 30 min. The anode is electrolyzed at 850 °C with a current density of 0.6 A/cm^2 for 72 h. The electrolyte is composed of cryolite + 20% excess $AlF_3 + 3\%Al_2O_3$. De Nora reported the results of electrolytic tests for 100 A, 1000 A, and 20,000 A cells. The amount of Fe in the aluminum product is less than 0.5%. Lowering the electrolysis temperature can reduce the Fe content, and the Ni in the Ni-Fe alloy hardly pollutes aluminum.

T. Altdorfer [7] reported a revolution in aluminum industry due to the development of inert anodes. Van leuwen [8] reported on the prospect of inert anodes. He believes that industrial-scale trials will be implemented in 2002 and will be applied in industry in 2005. Alcoa's Troutdale Aluminum Plant has been shut down. Inert anode batteries will be installed in 2003, and new anode batteries will be launched in the same year. Once the United States is successful, Peshne Co., France, will also adopt Alcoa's technology to develop inert anodes. The construction of a new plant with an inert anode can save 35% of the investment, and the electricity produced by aluminum is comparable to that of Hall-Heroult. It will take several years for the world aluminum industry to achieve large-scale technological transformation and the application of inert anodes.

Benedyk and De Nora proposed the following specifications for inert anodes:

1. When the current density is 0.8 A/cm^2, the corrosion rate of anode should be less than 10 mm/a.
2. When the current density is 0.8 A/cm^2, the polarization voltage should be less than 0.5 V. The connection voltage drop of inert anode should not be greater than that of the carbon anode.

In normal industrial production conditions, its performance should be fully satisfied with the following criteria:

1. Oxygen stabilization at 1000 °C
2. Resistance to fluorine corrosion
3. Thermal stability and appropriate thermal vibration resistance
4. Appropriate mechanical strength
5. Low resistivity
6. Without reducing the quality standard of aluminum
7. Easy and stable electrical connection
8. Being environmentally sound
9. Harmless to physical health and safety
10. Scope for technological improvement
11. Low cost and easy to make large pieces

These criteria are difficult to fully implement. According to Welch, inert anodes have the advantages of reducing anode replacement, eliminating costs related to the production of carbon anodes, and eliminating CO_2.

Alcoa and Moltech, a Swiss salt-soluble technology company, are the two largest companies in the world related to inert anodes. The longest electrolysis time is up to 6 months. Its main goal is to require the anode to consume as few as possible to ensure the quality of aluminum products. Although there are different opinions, it is necessary to judge which kind of anode is the most suitable based on the quality and the cost of aluminum.

It is difficult to meet all the above criteria. However, due to the unique and enormous advantages of inert anode, people can optimize the electrolyte system, the structure and process of the electrolytic cell, from the development of electrode materials [9], and the corresponding electrolyte system, technical and economic indicators assessment, and optimization [10].

The idea of inert anode electrolysis has been used for a long time. From the beginning of Hall-Heroult aluminum smelting process, C. M. Hall, the pioneer of electrolytic aluminum smelting, tried to adopt inert anode [11] in 1888. At first, Cu and other metal materials were chosen to form metal oxide layer on metal surface, so as to be used as inert anode materials. Later, people began to study oxide materials with low solubility and good semiconductor properties in cryolite melts. Then, Belyaev and Studentsov first tried to use various sintered oxides such as SnO_2, MO, and Fe_3O_4. In the 1930s, various inert anode materials (such as metals and alloys) and refractory hard metals (such as boride, carbide, and metal oxides) were extensively studied, and some progress has been made [12]. In 1981, K. Billehaug et al. [13] classified the inert anode materials into four categories: refractory hard metals (RHM), gaseous fuel anodes, metal anodes, and oxide anodes.

3.3 Detection and Testing of Inert Anodes

Only partially suitable materials for inert anodes have been reported, and more inert materials need to be carefully screened to determine. The following are the experimental methods for detecting whether the anode material is inert material.

3.3.1 Three-Electrode Cell Detection of Inert Anode

McLeod et al. [14] have studied the inert characterization of several kinds of anode materials, such as Pt, Au, nickel ferrate, cobalt ferrate, etc. The three-electrode test equipment used is shown in Fig. 3.2. The electrolyte composition used in the experiment is the same as that used in industry. The mass ratio of NaF/AlF_3 is 2.3, 5% CaF_2, and Al_2O_3 saturated by 11%. The temperature is 1000 °C.

Natural Greenland cryolite, reagent grade Al_2O_3, AlF_3 are supplied by Alcoa. The concentration of Al_2O_3 was analyzed by $AlCl_3$ dissolution method. In addition, the acidity of electrolyte was analyzed by thermal titration.

When the temperature reaches 960 °C and the electrolyte is stabilized, the electrolyte melts fully. The Pt anode is used for pre-electrolysis to eliminate impurities in the electrolyte. When the applied voltage reaches 3.00 V, the anode current density reaches 1 A/cm^2. The purity of the electrolyte, known as the residual current, is measured by voltage scanning (20 mV/s). Residual current is defined as the current when the voltage is slightly lower than the voltage required to restore Al_2O_3. When the electrolyte is turbid, the residual current is greater. When the electrolyte becomes clean, the residual current is less than 20 mA/cm^2. The pre-electrolysis usually takes 3 h.

After the pre-electrolysis is completed, the inert anode sample to be tested can be immersed in the molten solution for electrolysis, and then a slow steady voltage scan is performed. Their residual current values are all very small, as shown in Fig. 3.3. On the scanning diagrams of Pt and Au, the residual current is about

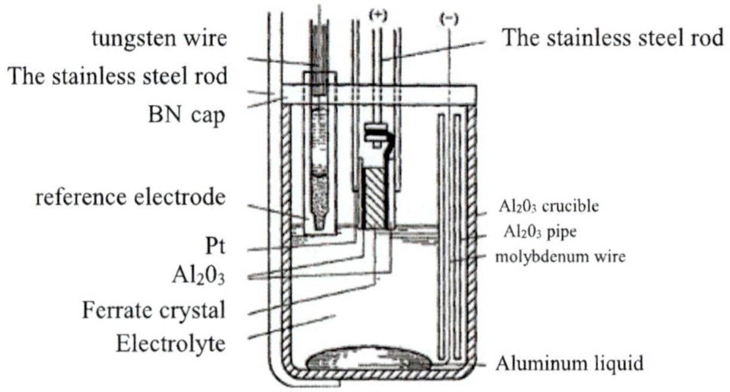

Fig. 3.2 Testing device for inert anode

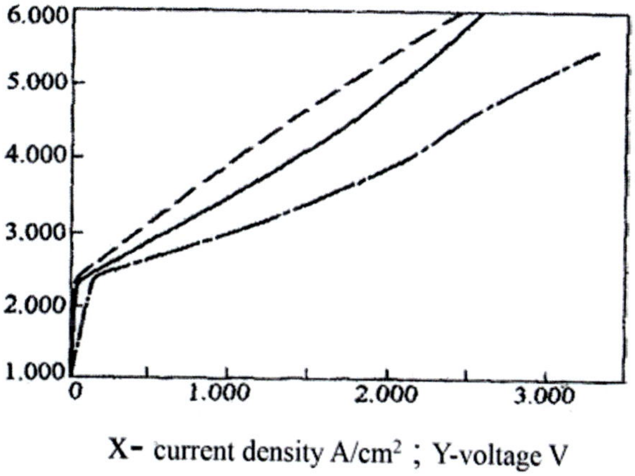

Fig. 3.3 The relationship between current density and anode voltage of platinum, gold, and cobalt ferrate (electrolyte composition NaF/AlF_3 = 2.2, CaF_2 5%, Al_2O_3 11%, using Greenland cryolite, 960 °C). *Solid line* Pt, *Dashed line* Au, *Dash-dotted line* Cobalt Ferrate. *X* current density (A/cm^2), *Y* over voltage (V)

20 mA/cm^2, indicating that the electrolyte is clean. Then, they are connected to the computer control and measurement device to achieve current pulses and faster voltage scanning. The value of the anode overvoltage is obtained by subtracting IR voltage drop and zero current voltage from the potential value between the anode and the reference electrode. The IR voltage drop is the voltage drop between the anode and the reference electrode, which is measured with the current breaker (ESC model 810-02). The zero current voltage is usually 2.16 V. The voltage from the anode to the reference electrode is extrapolated to the open circuit (i.e., $I = 0$). A total of 0.06 V voltage, including film voltage and thermoelectric potential, is included in the 2.16 V voltage. When the power is off, this part of voltage can also be calculated from the cathode aluminum liquid to the aluminum reference electrode.

At the end of the test, the solidified electrolyte and tin samples are taken out and the Ni, Fe, and Co content is analyzed.

Platinum, gold, nickel ferrate, and cobalt ferrate samples are determined by the above instruments and methods. The platinum sample has the best reproducibility. Therefore, we first test platinum sample. Three current-voltage curves, platinum, gold, and cobalt ferrate (contains 25% Co in Fe_2O_3), and their overvoltage values are shown in Fig. 3.4. These measurements are carried out under the steady current condition, and the results are satisfactory. If the three curves are extrapolated to zero current, the decomposition voltage of Al_2O_3 is 2.2 V. The residual current densities of both platinum and gold are less than 40 mA/cm^2. The residual current density of cobalt ferrate is 100 mA/cm^2. This results from the contamination of electrolytes by electrode materials during continuous use, as shown in Figs. 3.3 and 3.4.

3.3.2 *Inert Anode Electrolytic Test Cell*

Ray [15] made ceramic-based inert anode from Ni-Fe-O and Ni-Fe-Cu-O system materials and carried out electrolysis experiments on a small electrolytic cell in the laboratory. These materials have good electrical conductivity. In the electrolytic aluminum, the content of gold is low. The small electrolytic cell is shown in Fig. 3.5.

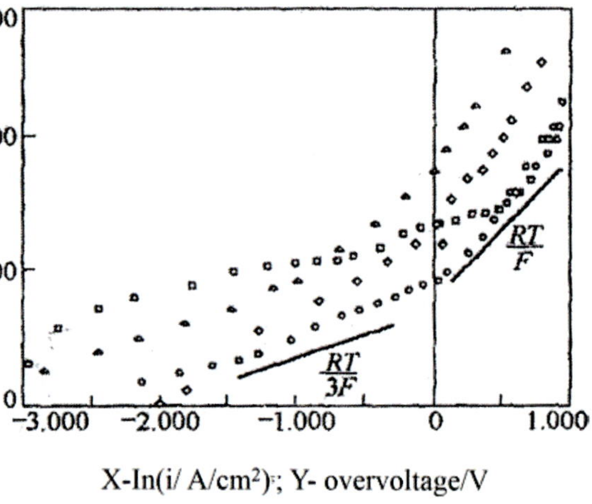

Fig. 3.4 The relationship between current density and anode overvoltage of platinum, gold, and cobalt ferrate (electrolyte composition NaF/AlF_3 = 2.2, CaF_2 5%, Al_2O_3 11%, using Greenland cryolite, 960 °C). *Open circle* Pt, *Open triangle* Pt, Thonstad measure 1000 °C, *Open square* Au, *Open diamond* Cobalt Ferrate

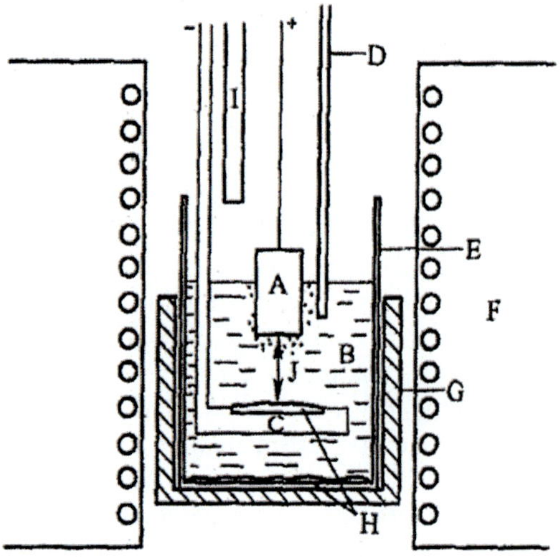

Fig. 3.5 Inert anode electrolytic test cell. *A* anode, *B* electrolyte, *C* cathode, *D* thermocouple, *E* alumina crucible, *F* electric furnaces, *G* protector, *H* liquid aluminum, *I* feeder, *J* pole distance

The results of the test cell are summarized in Table 3.4. The composition of the inert anode is 51.7% NiO and 48.3% Fe_2O_3 and compaction density of 5.1 g/cm^3, and electrolyte composition of NaF/AlF_3 is 2.2, CaF_2 of 5%, and Al_2O_3 of 5%.

Aluminum containing Fe of 0.15%–0.4% and Ni of 0.02%–0.09% can be obtained by using Ni_xFe_{3-x} and $Ni_xFe_{1-x}O$ oxide anode in aluminum electrolysis. However, the conductivity of the anode is only 0.8 Ω^{-1} cm^{-1} at 1000 °C. The corrosion rate of the anode is 2×10^{-4} cm/h, and the performance is not good. The addition of Cu can increase the conductivity of the electrolyte. The anode consumption decreases with the increase of Al_2O_3 content in the electrolyte.

Table 3.4 Test results of inert anode electrolysis composed of 51.7% NiO and 48.3% Fe_2O_3

Test number	Test time (h)	Total time (h)	Current efficiency (%)	Electrolyte analysis		Metal analysis		Remark
				ω(Ni)/%	ω(Fe)/%	ω(Ni)/%	ω(Fe)/%	
A-1	25		83	0.02	0.04	3.2	0.2	The anode does not corrode, the crucible is broken, and the Ni enters the aluminum
A-2	24	49	91	0.02	0.04	0.05	0.17	There is no obvious change in anode
B-1	24		96	0.03	0.02	0.02	0.09	No significant change in anode
B-2	29	53	89	0.03	0.04	0.36	0.40	Thermocouple may be eroded
B-3	26.5	79.5	91	0.02	0.03	0.09	0.23	Visible small size change
B-4	21	100.5	80	0.01	0.02	0.06	0.18	A small amount of corrosion, test total loss of poor connection 2.2×10^{-4} cm/h
C-1	24		91	0.02	0.03	0.05	0.15	No corrosion
C-2	25.5	49.5	83	0.02	0.03	0.06	0.27	No corrosion
C-3	27	76.5	81	0.02	0.03	0.03	0.19	Test total loss of 1.5×10^{-4} cm/h (at most)

3.3.3 *The Solubility of Inert Anode Materials in Cryolite Melts*

DeYoung [16] measured the solubility of several oxides in cryolite-based melts. The used device is shown in Fig. 3.6.

The reagents used in the test are Greenland cryolite, AlF_3 produced by Alcoa, reagent grade CaF_2 and calcined $Al(OH)_3$. After grinding, mixing, and analysis, the tablets were prepared by reagent grade NiO and Fe_2O_3 at a pressure of 179 Mpa. Then, it was sintered in air at 1350 °C. The density of the sample was 5.72 g/cm^3, Fe_2O_3 was 5.13 g/cm^3, and $NiFe_2O_4$ was 4.64 g/cm^3. In the experiment, the sample was broken into two to three pieces. The amount of added oxide was 2–3 g. Air was pumped to the surface of the electrolyte during the experiment and then bubbled into the electrolyte melt regularly. Air is pre-dried through columns of calcium sulfate.

At the end of bubbling, a sample of the electrolyte was taken out in 25 min. We analyzed the content of oxide and the ratio of substance to substance. Each experiment takes 80–100 h. The solubility was determined by adjusting the temperature

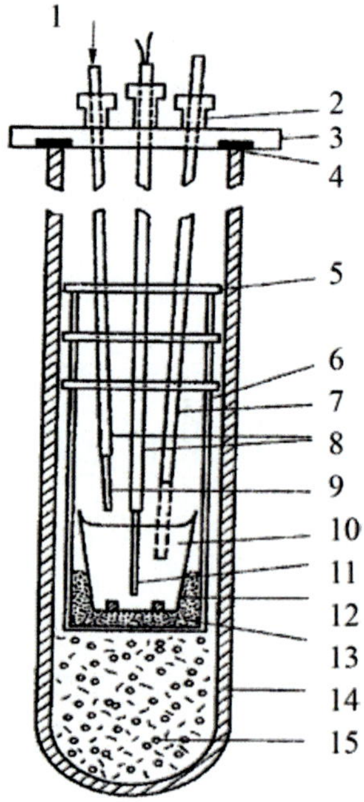

Fig. 3.6 Experimental device for measuring the solubility of oxide in cryolite melt. (*1*) Air, (*2*) Connection, (*3*) Copper cover, (*4*) Bed down, (*5*) Partitions, (*6*) Alumina crucible, (*7*) Corundum sampling tube, (*8*) alumina tube, (*9*) Pt tube, (*10*) Cryolite solution, (*11*) Thermocouple, (*12*) Pt crucible, (*13*) Particle Al_2O_3, (*14*) Alumina tube, (*15*) Foam Al_2O_3

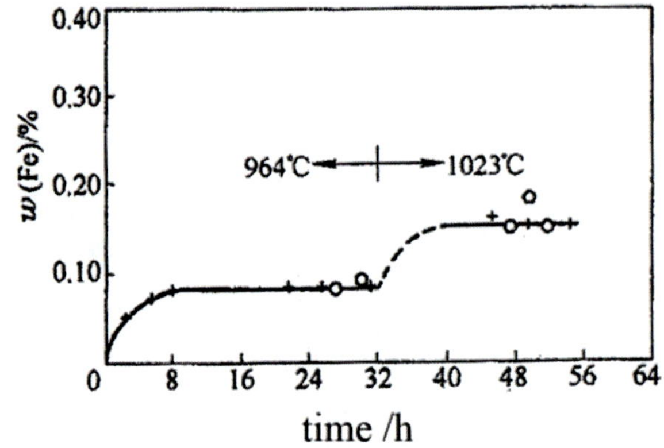

Fig. 3.7 Time for Fe_2O_3 to reach equilibrium with solution

every 24–30 h. The solubilities of Fe_2O_3 and NiO in cryolite melts are shown in Figs. 3.7 and 3.8, respectively. The composition of cryolite melt is as follows: the content ratio of NaF/AlF_3 is 2.0%, and the content of Al_2O_3 is 4.6%.

The solubility of Fe_2O_3, NiO, and $NiFe_2O_4$ in cryolite-alumina melt decreases with the increase of Al_2O_3 concentration in molten solution. The solubility of $NiFe_2O_4$ is lower than that of monomer Fe_2O_3 and NiO with the lower melt temperature. Therefore, low-temperature electrolysis should be combined with the inert anode.

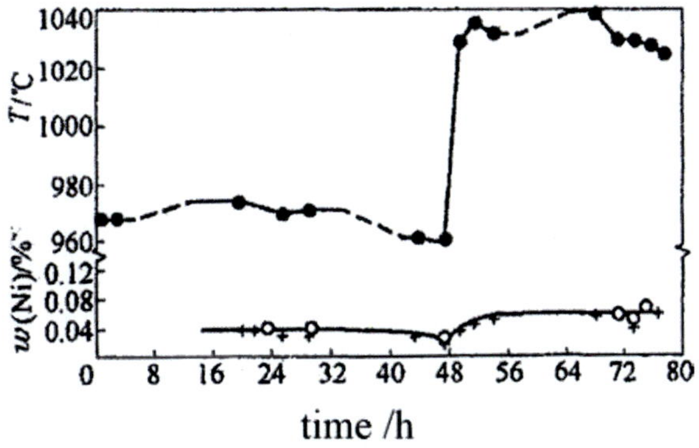

Fig. 3.8 Time for NiO to reach equilibrium with solution

3.3.4 Corrosion and Passivation of Ceramic Inert Anode During Electrolysis in Cryolite Melt

In recent years, the research on inert anode materials can be classified into two categories: (1) ceramic oxide materials and (2) metal alloy materials. The solubility of ceramic oxide materials in cryolite melts is low. However, the resistivity of them are high. Thus, only thin layer electrodes can be used. In addition, metal materials have good conductivity. However, in addition to precious metals and platinum, the oxide film formed on the surface is easy to collapse. The reason is that it cannot resist oxygen erosion in molten salt electrolysis. Therefore, neither of them can meet the requirements of inert anodes.

Tarcy [17] used two electrolytic cells to study the corrosion and passivation of inert anodes. The first type of electrolysis cell is a three-electrode system, which is used to study the corrosion of materials. The second kind of electrolytic waste is used to more strictly evaluate the most promising materials.

3.3.5 Corrosion Test Cell

The corrosion test cell is shown in Fig. 3.9.

In this three-electrode system, the detected anode and Pt wire are inserted from the top of the groove. The Mo wire is used as the auxiliary electrode; the aluminum encapsulated in the Al_2O_3 tube is used as the reference electrode. The tungsten wire is conductive.

To fix the working area of the anode, it is encapsulated with alumina tube. The anode was only 1 mm into the electrolyte. The ceramic material was subjected to constant current polarization (1 A/cm^2) for 3 h. The metal material was subjected to constant current polarization (1 A/cm^2) for 0.5 h. The linear potentiostatic scanning was performed. The scanning speed was 1 m/s, and the open circuit potential was

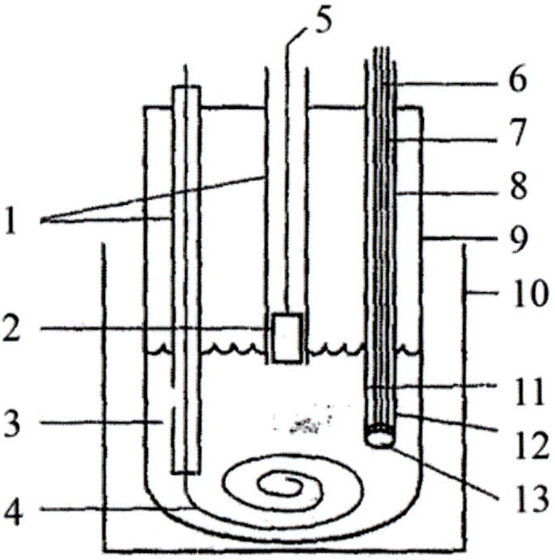

Fig. 3.9 Electrolysis cell for studying electrochemical corrosion of materials. (*1*) Al_2O_3 insulating tube, (*2*) inert anode, (*3*) electrolytes, (*4*) molybdenum auxiliary anode, (*5*) platinum wire, (*6*) tungsten wire, (*7*) alumina tube, (*8*) alumina cover, (*9*) alumina crucible, (*10*) inkang alloy containers, (*11*) capillary tube (1/16 in.), (*12*) reference electrode, (*13*) aluminum

taken as the zero-current potential. The corrosion current was used as the anode current. This is produced when the potential of O_2 precipitation is lower than that of thermodynamics. After the electrolysis was finished, the anode was cut open for electron probe analysis.

The suitability of a material for an inert anode depends on the following three conditions: (1) the corrosion current is small; (2) the circuit potential is high; and (3) the gold phase microstructure in the anode is integral.

The electrolyte composition used in short-time electrolysis is as follows: the volume ratio of NaF/AlF substance is 2.3/CaF_2 5%, and the Al_2O_3 concentration of electrolyte is saturated and the temperature is 980 °C. Electrolytes used in long periods of electrolysis are the same. The difference is that the consumed alumina must be replenished regularly (based on 100% current efficiency).

3.3.6 Test Cell for Long-Time Electrolysis

The test cell for long-time electrolysis is shown in Fig. 3.10.

Ceramic materials are prepared by reactive sintering or traditional sintering. The following reaction occurs during the reactive sintering process:

$$5NiO + Fe_2O_3 + 2Fe — 2NiFe_2O_4 + 3Ni \tag{3.6}$$

In the above two tests, the specimen was pressed by isostatic pressure 14 l Mpa.

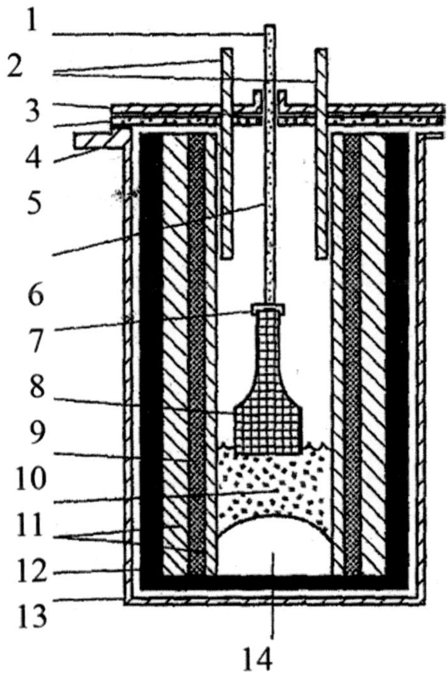

Fig. 3.10 Electrolytic cells for long-term testing of inert anodes: (*1*) anode connecting rod, (*2*) alumina input tube, (*3*) inconel alloy cover, (*4*) keep warm cushion, (*5*) cathode connection, (*6*) Ni connecting rod, (*7*) electromagnetic mechanical support, (*8*) inert anode, (*9*) Al_2O_3 tube, (*10*) electrolytes, (*11*) Al_2O_3 lining, (*12*) carbon crucible, (*13*) inkang alloy crucible, (*14*) aluminum liquid

3.3.7 Newly Developed Inert Anode

The research of general inert anode is mainly focused on cermet, especially Fe-Ni-Cu-O system. Almost all studied cermet are based on metal oxide ceramics. In addition, a small amount of metal phase (15–22%) dispersed in the ceramic phase. However, such anode materials have obvious inherent ceramic defects.

3.4 Research Progress of Inert Anodes in Recent Years

After the 1980s, the research work of inert anode materials has mainly been focused on the development and testing of metal oxide ceramic anodes, alloy anodes and cermet anodes. Therefore, we mainly introduce the latest research progress of these three kinds of inert anodes in recent years.

Table 3.5 Solubility of some oxides in Na_3AlF_6 and Na_3AlF_6–Al_2O_3 solution at 1000 °C [18]

Oxide	Solubility in Na_3AlF_6 (mass fraction) %	Solubility in Na_3AlF_6–5% Al_2O_3 (mass fraction) %
Na_2O	23.0	–
K_2O	28.0	–
BeO	8.95	6.43
MgO	11.65	7.02
CaO	16.3	–
BaO	35.75	22.34
ZnO	0.51	0.004
CdO	0.98	0.26
FeO	6.0	–
CuO	1.13	0.68
NiO	0.32	0.18
Co_3O_4	0.24	0.14
Mn_3O_4	2.19	1.22
B_2O_3	Infinite	Infinite
Cr_2O_3	0.13	0.05
Fe_2O_3	0.18	0.003
Le_2O_3	18.8 (1030 °C)	19
Nd_2O_3	21.3 (1050 °C)	–
Sm_2O_3	20.4 (1050 °C)	–
Pr_6O_{11}	31.4 (1050 °C)	–
SnO_2	8.82	–
TiO_2	5.91 (1030 °C)	3.75
SnO_2	0.08	0.01
CeO	16.1	–
V_2O_5	1.20 (1030 °C)	0.65
Ta_2O_3	0.38	–
WO_3	87.72	86.14

3.4.1 Metal Oxide Ceramic Anode

Compared with other alternative materials, metal oxide ceramics have the advantages of low solubility and low corrosion rate in electrolyte melt. The solubility data of various oxides in aluminum electrolyte melt are shown in Tables 3.5, 3.6, and 3.7 [18–20]. Keller et al. [21] considered that the lifetime of metal oxide ceramic anode strongly depends on the dissolution rate of the electrode component in the electrolyte. The dissolution rate mainly depends on the reduction of the anode component near the cathode. However, its development is limited by its poor conductivity at high temperature, thermal shock resistance, and machining performance. Metal oxide ceramic anode materials can be classified into composite metal oxides, single metal oxide, and metal oxide mixtures.

Table 3.6 Solubility of some oxides in Na_3AlF_6 and Na_3AlF_6–Al_2O_3 solution at 1100 °C [19]

Oxide	Solubility in Na_3AlF_6 (mass fraction) %	Solubility in Na_3AlF–5% Al_2O_3 (mass fraction) %	Solubility in Na_3AlF_6–Al_2O_3 (saturated solution) (mass fraction) %
Cu_2O	0.28	0.23	0.34
ZnO	2.9	0.17	0.025
FeO	5.4	3.0	0.30
NiO	0.41	0.09	0.0076
CuO	1.1	0.44	0.56
Co_3O_4	7.3	–	–
Cr_2O_3	0.70		
Fe_2O_3	0.8	0.4	0.22
TiO_2	5.2	–	4.54
ZrO_2	3.2	–	–
SnO_2	0.05	0.015	0.01
CeO_2	3.4	1.0	0.6

Table 3.7 Solubility of some oxides in Na_3AlF_6 and Na_3AlF_6–Al_2O_3 solution at 1100 °C [20]

Oxide	Solubility (mass fraction) %
$MgCr_2O_3$	Mg 1.06; Cr 0.04
$CoCr_2O_3$	Co 0.01; Cr 0.01
$NiFe_2O_4$	Ni 0.02, 0.009
	Fe 0.05, 0.058
$ZnFe_2O_4$	Zn 0.01; Fe 0.04
$LaCoO_3$	La > 1.0; Co 0.14
$SnCo_2O_4$	Sn 0.02
	Co 0.01

3.4.2 *Spinel (AB_2O_4) Composite Metal Oxide Anodes*

Spinel composite oxide ceramics have been widely studied as an alternative material for inert anodes due to their good thermal stability and favorable electrocatalytic activity (low overpotential) for oxygen evolution. Among them, the spinel complex oxides, such as $NiFe_2O_4$, $CoFe_2O_4$, $NiAl_2O_4$, $ZnFe_2O_4$, and $FeAl_2O_4$, have been studied. The corrosion behavior of ferrate of Ni and CO was studied by Augustin et al. [22]. The results show that spinel oxide ceramics have good corrosion resistance in cryolite molten salt electrolytes. The corrosion resistance of $ZnFe_2O_4$ [23] has been studied. The corrosion rate of $ZnFe_2O_4$ is greatest at the current density of 0.5–0.75 A/cm^2.

In 2001, Galasiu et al. prepared $NiFe_2O_4$ ceramic materials by "coprecipitation sintering" [24]. The results show that the properties of inert anode prepared by this process are better than that of conventional "solid-phase synthesis-sintering" and reactive sintering methods. Y. Zhang et al. [25] proposed a dissolution model for $NiO/NiAl_2O_4$ and $FeO/FeAl_2O_4$ in cryolite melts. The former assumes that Ni exists in the form of two complex ions after dissolution, and the latter assumes that FeF_2,

Na_2FeF_4, and Na_4FeF_6 exist. The experimental results show that these hypotheses are in good agreement with the experimental data. In addition, Julsrud et al. [26] carried out the electrolysis experiment of $NiFeGrO_4$ anode materials in 2001, and the anodic arrangement in aluminum reduction cell was put forward.

3.4.3 SnO_2-Based Metal Oxide Anodes

SnO_2-based anodes has been used as the preferred material for inert anodes by many researchers. Yang Jianhong et al. [27] studied the behavior of SnO_2-based anode in aluminum electrolytes. The oxygen evolution overpotential of SnO_2-based anode in electrolyte containing 10% Al_2O_3 was measured at a molar ratio of 2.7. The results show that the anode doped with trace amount of Ru, Fe, and Cr has significant electrocatalytic effect. Qiu Zhuxian et al. [28] studied the effect of ZnO, CuO, Fe_2O_3, Sb_2O_3, Bi_2O_3, and other oxide additives on the formation and conductivity of SnO_2-based anodes and carried out 100-year electrolysis test.

Haarberg et al. [29] found that the solubility of SnO_2 in cryolite melt at 1035 °C is 0.08%. The solubility is higher under reductive conditions, such as carbon slag and dissolved aluminum in the electrolyte. They believed that the increased solubility of SnO_2 is due to the presence of Sn^{2+} or Sn^{+} in the electrolysis process, and the dissolved Tin ions are reduced to Tin metal at the cathode. Issaeva [30] and Yang Jianhong [31] tested the electrochemical properties of SnO_2. Their cyclic voltammetry measurements were carried out using Pt, Au, and glass-like C as working electrodes. The voltage curves show that the peak value is related to the two oxidation states of Tin (such as Sn^{2+}, Sn^{4+}) in the molten salt. In the molten salt without other oxides, the volatiles of SnF_2 and SnF_4 are found on the anode. If there is dissolved alumina, it will form a stable substance with the dissolved Sn instead of volatile substances.

In 1996, Sadus et al. [32] studied the behavior of SnO_2-based inert anodes doped with 2% Sb_2O_3 and 2% CuO in different electrolytes. They measured the corrosion rate of SnO_2-based anode at different temperatures. Through SEM and EDS analysis of anode samples, the copper elements in the anode have loss, and aluminum-rich layer appears on the anode surface under certain conditions. Popescu et al. [33] have measured the current efficiency, electrolysis temperature, current density, and polar distance of the anode with the same composition studied by Sadus under laboratory conditions. The electrolyte composition and properties during the anodic effect are discussed. Galasiu [34] studied the effect of Ag_2O on the electrochemical properties of SnO_2 anodes. The results show that the resistance of the anode is the lowest and the corrosion resistance is the best when the anode composition is 96% SnO_2, 2% Sb_2O_3, and 2% Ag_2O. Las [35] studied the effect of tantalum, niobium, and antimony on the electrical conductivity of ceramics.

In 2000, Cassyre et al. [36] used a transparent electrolyzer to study the anodic gas generation process when SnO_2 was used as the anode. It is further confirmed that the surface of the anode has good wettability with the electrolyte when the inert anode is used.

3.4.4 CeO_2-Coated Anode

The patent filed by Eltech Systems in 1986 [37] pointed out that when trivalent Ce is dissolved in an aluminum electrolyte, the so-called CER0X coating composed of Ce^{4+} oxyfluoride compounds can be deposited on the anode surface under electrolytic conditions. Walker et al. [38] reported that CER0X can reduce the corrosion of SnO_2 anode matrix.

The solubility test shows that CeF_3, Ce_2O_3, and CeF_4 can be dissolved in the aluminum electrolyte melt. However, the solubility of CeO_2 is low. When Ce^{3+} is added to the aluminum electrolyte melt, it can be reacted according to Formula (3.7):

$$CeF_3 + \frac{1}{2}Al_2O_3 + \frac{1}{4}O_2 = CeO_2 + AlF_3 \tag{3.7}$$

In order to maintain the stability of CEROX coating on anode surface, it is necessary to maintain a certain concentration of CeF_2 and Al_2O_3 in the electrolyte melt.

Although CEROX can reduce the corrosion of anode matrix, there are three problems in practical applications. First, the CeF_3 in the melt can not only undergo anodic oxidation deposition but also be reduced at the cathode according to Formula (3.8):

$$CeF_3 + Al = Ce_{\text{in Al}} + AlF_3 \tag{3.8}$$

Ce entering cathodic aluminum liquid contaminates cathode products. Therefore, it is necessary to remove Ce into molten aluminum and recycle it back into electrolyte [39]. Secondly, the CEROX coating is not very dense, and the substrate corrosion occurs during electrolysis. This causes the coating to flake off. In addition, it is necessary to control the thickness of CEROX coating effectively to ensure that the anode has higher conductivity. This is difficult in practical operation.

In 1993, J. S. Gregg et al. [40] used CeO_2 coating as an inert anode for aluminum electrolysis using $NiFe_2O_4^+$, 18% NiO^+, and 17% Cu cermet as the substrate. The corrosion resistance of the anode is improved, and the corrosion performance is closely related to the content of CeO_2 in the coating. After a long period of electrolysis, the inert anode coated with CeO_2 still has corrosion cracks. Yang Jianhong et al. [27] studied the inert anode coated with CeO_2 on SnO_2 substrate. The conductivity of inert anodes coated with CeO_2 increases, and the corrosion resistance of inert anode based on SnO_2 is enhanced. In addition, the wettability is better between SnO_2-based inert anode with CeO_2 coating and the electrolyte. In 1995, E.W. Dewing et al. [41] studied the dissolution of CeO_2 in cryolite molten salt. The dissolution of CeO_2 is related to the partial pressure of oxygen, the content of aluminum, and aluminum fluoride in molten salt. Ce exists mainly in the form of Ce^{3+} in molten salt. The main product after condensation is CeF_3.

3.4.5 Other Metal Oxide Electrodes

In addition to the metal oxide ceramic anodes, some inert anodes of metal oxide ceramic reported by the patent are shown in Table 3.8.

In 1999, the inert anode with 62.3% Cr_2O_3, 35.7% NiO, and 2% CuO was electrolyzed by Pietrzyk et al. [43]. The results show that the impurity content of aluminum is less than 0.3% when the corrosion rate of the anode is less than 1 cm/A.

In 1995, Zaikov et al. [44] measured the anodization with NiO–2.5% Li_2O in alumina-saturated electrolyte for 4.5 h. The obtained anode was observed in good

Table 3.8 Metal oxide ceramics proposed as inert anode for aluminum electrolysis [42]

Material composition	Resistivity 1000 °C/Ω cm)	Structure and type of conduction	Preparation process
1. 98% SnO_2 + 1.5% Sb_2O_3+0.3% Fe_2O_3 + 0.2% ZnO	0.1–10	Rutile N type semiconductor	1350–1450 °C Sintering 15–20 h
2. 96% SnO_2 + 2% CuO + 2% Sb_2O_3	0.004	Rutile N type semiconductor	1350 °C Sintering 2 h
3. 65% Y_2O_3 + 15% Ti_2O_3 + 20% Rh_2O_3	5	–	1200 °C Sintering 5 h
4. $CoCr_2O_3$ (62.3% Cr_2O_3 + 35.7% Co + 2% NiO)	1	Spinel	1800 °C Sintering 2 h
5. $LaCrO_3$ (60.2% La_2O_3 + 33.9%Cr_2O_3 + 5.9%$SrCO_3$	0.1	Perovskite	1900 °C Sintering 1 h
6. $LaNiO_3$ (65.8% La_2O_3+ on titanium 33.7% Ni_2O_3 + 0.5% In_2O_3	1	Perovskite	Plasma spraying Substrate after preheating
7. $PdCoO_2$ + $PtCoO_2$ (55.4% PdO + 5% PtO + 39.6% CoO)	0.01	Delafossite	900 °C Sintering 24 h
8. $ZrCeO_4$ + $ZrSnO_4$ (44.4% ZrO_2 + 3.7% CeO_2 + 48.9% SnO_2 + 2% CuO + 1% non-oxide	1	Scheelite	Plasma spraying on platinum plated titanium substrate after preheating
9. $Ni_{0.6}Sn_{0.4}Fe_{1.2}Ni_{0.1}O_4$	0.2	Spinel	1400 °C Sintering 24 h
10. $Ni_xFe_{1-x}O$ + FeNi and $NiO.NiFe_2O_4$	–	Spinel	–
11. Ba $Ni_2Fe_{15.54}Sb_{0.16}O_{27}$ + 16% (bulk) metal	–	–	–

condition. During the experiment, the corrosion rate of the anode was calculated by weighing the mass of the anode before and after electrolysis. The results show that the corrosion rate of the oxide electrode depends on its preparation parameters. Prolonging the sintering time and increasing the sintering temperature are beneficial to reducing the corrosion rate.

3.5 Study on Alloy Anode

In the 1990s, many researches have been conducted on inert anode materials. The alloy anode has the advantages of high strength, no embrittlement, good electrical conductivity, strong thermal shock resistance, easy processing and manufacture, and easy connection with metal guide rod. Sadoway [45] believed that alloys are the best alternative materials for inert anodes. However, since the metal is highly active and unstable under high-temperature oxidation, a uniform, dense, and self-repairing protective film can be formed on the anode surface of the alloy. It is important to balance the dissolution rate and the forming rate of the film. This is also the main obstacle to the research and development of alloy anodes. The research progress alloy anode in recent years is as follows.

3.5.1 *Cu-Al Alloy Anode*

In 1999, J. N. Hryn and M. J. Pelin et al. [46] proposed that one of the components may be the "dynamic alloy anode" of Cu and (5–15%) A1, as shown in Fig. 3.11. It is a cup-shaped Cu-Al alloy container filled with molten salts with molten aluminum. It is transported through the wall of the alloy to the surface of the container

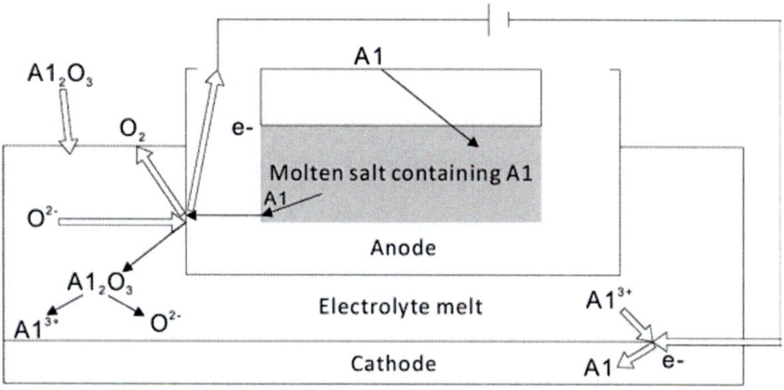

Fig. 3.11 Schematic diagram of "dynamic metal anode"

and oxidized by anodic electrochemistry (or anodic gas) to form a dense Al_2O_3 passivation film. Thus, the matrix alloy can be protected from oxidation and corrosion.

The Al_2O_3 passivation film continuously dissolves under the action of electrolyte. The regeneration and supplement of Al_2O_3 passivation film can be realized by the diffusion and oxidation of aluminum in molten salt. When the dissolution rate of the Al_2O_3 passivating film is equal to the diffusion replenishment rate, the Al_2O_3 film can exist stably in a certain thickness. In addition, the oxidation and corrosion of the anode matrix are avoided. In addition to the above structures, many studies have been carried out to reduce the corrosion of the Al_2O_3 passivating film by using low-temperature electrolytes with plate-shaped or rod-shaped Cu-Al alloys inert anodes [47, 48].

3.5.2 Ni-Fe-Based Alloy Anode

In 1994, T. R. Beck [49] explored the low-temperature aluminum electrolyte for Ni-Fe-Cu alloy anode (composed of 15Fe-70Cu-35Ni or 13Fe-50Cu-37Ni). The used electrolytes are composed of NaF-AlF_3 or NaF-KF-LiF-AlF_3 electrolytic at a temperature of 750 °C. The experimental results show that the corrosion rate of alloy anode in electrolysis is low, which is similar to the oxidation rate of alloy in air at the same temperature (750–800 °C). The results show that the prospect of low-temperature aluminum electrolysis with alloy anode is attractive.

In 1998, J. A. Sekhar et al. [50] used Ni-AI-Cu-Fe alloy as anodes. The optimal composition of the alloy is Ni-6Al-10Cu-11Fe-3Zn. The disadvantages of the alloy are that the oxidation rate is high and the anode surface is easy to be damaged during electrolysis. Thus, the corrosion resistance is poor. However, the oxidation rate can be slowed down by adding a small number of additives such as Si, Ti, and Sn to the Ni-AI-Cu-Fe alloy. It is pointed out that how to slow the oxidation rate of the alloy is the research focus of inert anode.

Duruz et al. [51] proposed a conductive layer coated on the alloy (such as Ni-Fe alloy) in 1999. First, the conductive layer does not allow atomic oxygen and oxygen molecules to permeate. Thus, it can protect the alloy. Second, it has a certain electrochemical activity, which can make the anodic oxidation of oxygen-containing complex ions take place at the anodic/electrolyte interface into new ecological oxygen atoms and ensure the smooth progress of the anodic reaction. In order to improve the surface corrosion resistance of high metal anodes, Duruz et al. proposed a nickel-rich Ni-Fe anode, which is composed of Ni-30% Fe alloy. After pre-oxidation in air at 1100 °C of 30 min, the alloy was electrolyzed at current density of 0.6 A/cm^2 and electrolysis temperature of 850 °C for 72 h. The electrolyte was 77% Na, AlF_6 20%, and AlF_3 3% Al_2O_3.

From 1998 to 2004, with the aid of the US Department of Energy, Northwest Aluminum Technology Company used Cu-Ni-Fe alloy inert anode and TiB_2 wettable cathode to suspend alumina particles in vertical electric tank in supersaturated electrolyte melts. The electrolysis experiments of low temperature (740–760 °C) for 300 h were carried out. The current efficiency reached 94%, and the purity of primary aluminum reached 99.9% (only the impurity elements introduced by anodic corrosion are considered). On this basis, a further 5000 A electrolytic test was prepared. Figure 3.12 is a photo of the welded anode (i.e., the electrolytic cell) [52].

On the basis of its previous research and patent technology, Moltech Company has developed Fe-Ni-based alloy inert anode of veronica, in which Cu, AI, Ti, Y, Mn, Si, etc. were added. The addition of these elements helps the alloy to form a dense and uniform surface passivation film after heat treatment and electrolysis. This can prevent the oxidation and corrosion of grain boundaries, thus improving the oxidation resistance and corrosion resistance of the alloy [53]. Moltech develops de Nora anode based on Veronica anode. By electroplating Co-Ni alloy coating in $NiSO_4$ and $CoSO_4$ solution, the active semiconductor coating of $Ni_xCo_{1-x}O$ is formed after oxidation in the air at 920 °C, which makes the anode have good electrochemical activity (lower overpotential) and conductive property. In the subsequent 100–300 A electrolytic test, the oxidation rate of the alloy substrate is 2 mm/A under steady state, the dissolution rate of the oxide coating is 3 mm/A, and the life of the extrapolated anode is over 1 year. In addition, the anodic component content in the primary aluminum is less than 0.1% [54].

On this basis, Moltech systematically studied the casting process, shape structure (Fig. 3.13), physical and chemical properties (Table 3.9), oxygen evolution potential (Fig. 3.14), electrothermal field, electromagnetic field, flow field of aluminum solution of de Nora alloy anode, electrolyte flow field under anodic bubble disturbance [56], new electrolytic cell structure, etc. [55].

Laboratory electrolysis test (<300 A) and expanded scale electrolysis test (4 kA and 25 kA) were carried out in different scales. The related technical and economic indexes were evaluated, and the technical prototype that was considered to be available for industrial test was put forward [57].

Fig. 3.12 5000 A electrolytic cell and alloy anode structure constructed by Northwest Aluminum Technology Corporation [52]

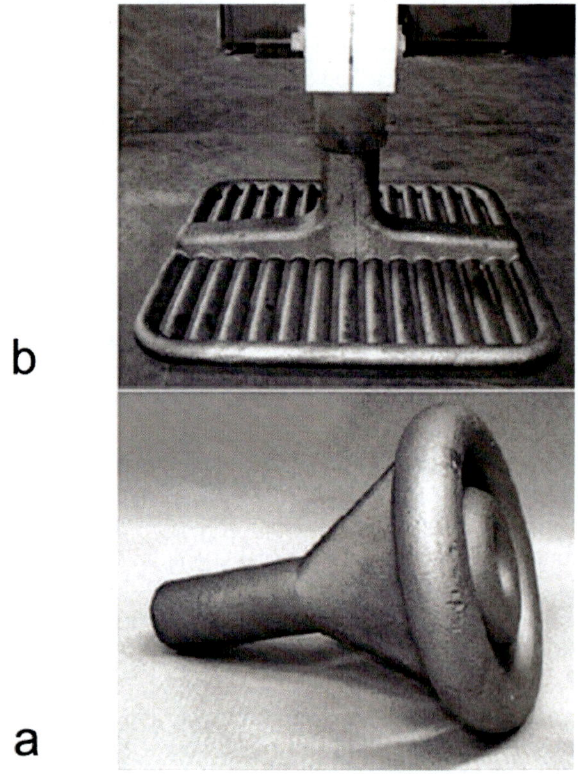

Fig. 3.13 The node structure of de Nora alloy in Moltech [54]. (a) Φ120 mm anode; (b) 600 mm × 600 mm anode

Table 3.9 Physical and chemical properties of de Nora anode [55]

Project	Index	Project	Index
Alloy matrix resistivity/Ω cm	3×10^{-7}	Oxide coating dissolution rate	2.9×10^{-3}
Oxide coating		Linear deformation	2.0×10^{-3}
Resistivity/Ω cm	3×10^{-2}	m/a	
Overpotential/V	0.10		
Oxidation rate of Alloy matrix/m/a	1.8×10^{-3}	Expected life/a	1–1.5

3.6 Study on Cermet Anode

Cermet is a composite composed of metal or alloy and ceramics. Metals and ceramics have their own advantages and disadvantages. Both metals and alloys have good mechanical properties and good electrical conductivity. However, their thermal stability and corrosion resistance are poor. They are easy to be oxidized and creep at high temperatures. Ceramics have high brittleness, poor electrical conductivity, good thermal stability, high fire resistance, and strong corrosion resistance. Cermet

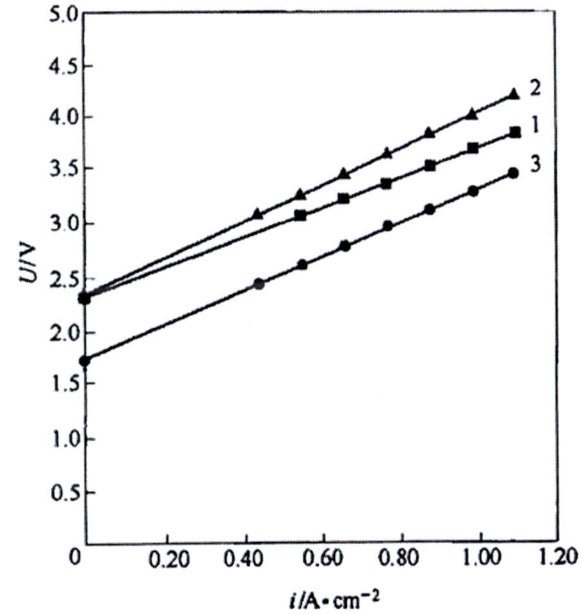

Fig. 3.14 The potential of various anodes in Na_3AlF_6 melts at 930 °C [54]. (*1*) de nora alloy anode, (*2*) $NiFe_2O_4$-coated alloy anode, (*3*) carbon anode

is a combination of metal and ceramics with the advantages of high hardness, strength, corrosion resistance, wear resistance, temperature resistance, good mechanical properties, and electrical conductivity. The ideal cermet anode has not only the strong corrosion resistance of the metal oxide ceramic anode but also the good electrical conductivity and mechanical properties of the metal anode. This can overcome the poor thermal shock resistance of the metal oxide anode and the difficulty of connecting with the anode guide rod, etc. It also has better corrosion resistance and oxidation resistance than metal or alloy anodes. At present, the inert anode of cermet generally takes oxide ceramics as continuous phase to form corrosion resistance, oxidation resistance network, and metal phase dispersion to improve the mechanical properties and conductive properties of materials. The corrosion resistance should also be considered in the selection of metal phase. Generally, the metal or alloy with oxide protective layer can be formed on the surface of the electrode under anodic polarization. Thus, the electrode has better corrosion resistance. However, since the metal oxide ceramics and metals cannot compensate for each other's disadvantages, it is difficult for the fabricated cermet materials to have the advantages of both metallic and ceramic phases. Some even introduced their own disadvantages. This is an important subject to be solved in the research of cermet inert anode materials. In recent years, the main research contents of cermet anode are as follows.

3.6.1 $NiFe_2O_4$-Based Cermet

In order to develop, prepare, and evaluate different inert anode materials, the American Aluminum Company (Alcoa) carried out systematic research on $NiFe_2O_4$ cermet inert anode from 1980 to 1985 with the aid of DOE. A research report and academic paper on cermet inert anode materials were published in 1986 [58, 59]. The Alcoa report confirmed that when the raw ingredient is 17% Cu, 42.91% NiO, and 40.09% Fe_2O_3, the $NiFe_2O_4$-based cermet (known as "5324-17Cu" cermet) have the best properties. The conductivity of the electrode is 90 s/cm. After 30 h electrolysis, the electrode shape is almost unchanged, showing good corrosion resistance and conductivity in small-scale test. Since then, $NiFe_2O_4$-based cermet has become the most important inert anode material for aluminum electrolysis and has been widely studied.

With the aid of the Department of Energy, Pacific Northwest National Laboratory (PNNL), a national laboratory in the Northwest Pacific, USA, has continued to study the inert anode of $NiFe_2O_4$-NiO-Cu cermet based on the results of Alcoa research since 1985. At the same time, Eltech Research has developed a CEROX coating for cermet inert anodes, funded by the US Department of Energy. Since 1991, PNNL and Eltech have carried out 6 kA electrolysis test of cermet inert anode in cooperation with Reynolds Metals Company, respectively [60]. As shown in Fig. 3.15, after 25 days of electrolytic test, the main problems are the poor thermal shock resistance of the large size anode, the electrode cracking, the serious damage of the conducting rod, and the uneven distribution of the anode current. The increase in cathode voltage is due to the formation of alumina precipitate at the bottom of the cell [61]. The corrosion resistance of cermet anode was greatly enhanced by using CER0X coating and was closely related to the content of CeO_2 in the coating. After a long period of electrolysis, cracks appeared in the inert anode coated with a CeO_2

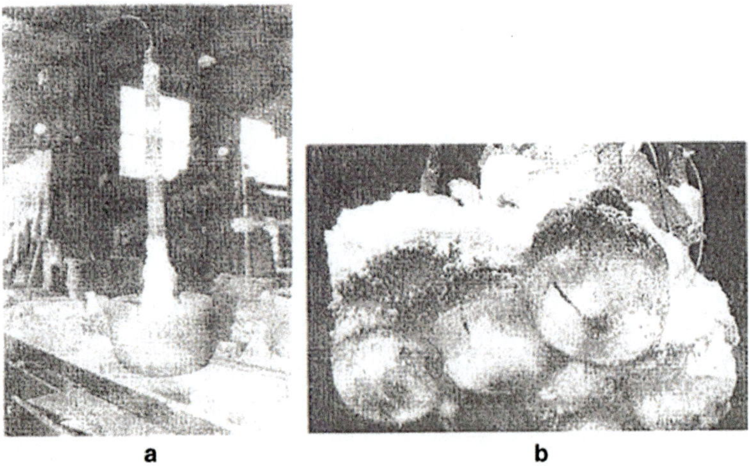

Fig. 3.15 Inert anode for 6 kA electrolytic test of Reynolds Metals Company [60]. (**a**) Flowerpot inert anode before electrolysis; (**b**) flowerpot inert anode after electrolysis

layer, and there was high Ce content in primary aluminum [62, 63]. Although the above experiments failed, the study of $NiFe_2O_4$-NiO-Cu cermet inert anode has been carried out. Alcoa published technical reports in 1997, 1999, and 2001. These reports indicated that the above problems have been solved to a certain extent [64–66]. Ray [67] reported in 2001 patents that the impurity contents of the primary aluminum obtained by anode electrolysis with $NiFe_2O_4$ + NiO + Cu are 0.2% Fe, 0.1% Cu, and 0.034% NiO. In 2001, Blinov [68] used the anodic composition of 65% $NiFe_2O_4$–18% NiO–17% Cu for electrolysis at a low temperature of 800 °C. The annual corrosion rate of the inert anode is 1.4 cm. Lorentsen et al. [69] also studied the impurity migration mechanism caused by this inert anode material between anode and cathode. In 1997, Blinov et al. [70] carried out low-temperature aluminum electrolysis experiments on inert anodes. The anode is $NiFe_2O_4$ + 18% NiO + 17% Cu, which is provided by Alcoa. The electrolysis temperature is 800 °C, and the anode current density is 0.2 A/cm^2. After 130 h of electrolysis test, the corrosion rate of anode is less than 10^{-3} g/(cm^2 h). The corrosion rate of the same anode at 950 °C is greater than 8×10^{-3} g/(cm^2 h) under this condition. This study shows that low-temperature electrolysis has the positive significance in reducing the corrosion rate of inert anodes.

To improve the electrical conductivity of cermet, Alcoa [71] added Ag to the $NiFe_2O_4$ + NiO + Cu cermet. The oxides of nickel and iron in cermet are about 50%–90%. The content of copper and silver or copper-silver alloy can reach 30%. The copper and silver alloy contains 90% copper and 10% silver. The results show that lowering the temperature can improve the corrosion resistance of the electrode. The mole ratio of electrolytes is 0.8–1.0 (contain 6% CaF_2 and 0.25% MgF_2), and the optimal electrolytic temperature is 920 °C.

In September 2001, Alcoa conducted a small-scale industrial trial at a smelter in Italy. It wished to build an industrial-scale electrolyzer in the United States that was operated entirely with inert anodes. At that time, Alcoa planned to increase its manufacturing capacity of inert anode to the level of one inert anode per day. They established the first production series of inert anode electrolysis cell in 2002. According to the report to the US Department of Energy, Alcoa plans to replace inert anodes in its carbon anode electrolyzer within 2 to 3 years [72]. However, Alcoa later announced that they had delayed the deployment of the next phase of research on inert anodes. The reason is that the thermal brittleness of the material and the connection with the guide rod have not been well solved [73].

3.6.2 *Experimental Study on Electrolytic Cell of Cermet Inert Anode at Initial Stage*

At the beginning of the development of inert anode, Alcoa considered $NiFe_2O_4$/NiO-Ni cermet as the most promising candidate. Because $NiFe_2O_4$ and NiO provide the anode with strong shape stability, the solubility in cryolite melt is low. Metallic

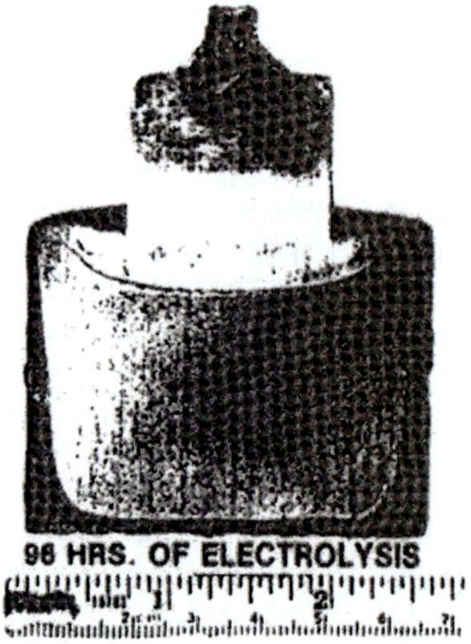

Fig. 3.16 $NiFe_2O_4$ NiO-Ni inert anode

Ni provides high electrical conductivity and mechanical strength. However, there is a problem of alloy phase preferential corrosion in this material system.

After 96 h of electrolysis, the black part of the anode indicates that it has deteriorated (Fig. 3.16), the conductivity of this material is only 0.5 Ω^{-1} cm^{-1}, and the conductivity of raw material is 300 Ω^{-1} cm^{-1}. Its mechanical strength is also low. The important difference from the remaining part is that the metallic phase is almost completely lost, and the solidified electrolytic mass is filled in the hole.

This is due to anodic selective dissolution of nickel:

$$Ni = Ni^{2+} + 2e \tag{3.9}$$

The anodic dissolution process shows that it produces anodic corrosion current under the potential of precipitated O_2. Using $NiFe_2O_4$ NiO-Ni alloy ceramic anode and Pt as working electrodes, the linear voltammetry is shown in Fig. 3.17. On the ceramic anode, the anode current appears to be lower than the potential of O_2 thermodynamic precipitation, and there is no anodic current on the platinum anode. The voltage to produce O_2 on the anode is between 2.1 and 2.2 V from the voltammetry. The results of electrochemical measurements of several metal and alloy materials are listed in Table 3.12.

Pt is the best inert anode material. However, it is easy to embrittle, as shown in Table 3.10. Cu is also better. However, it is easy to be oxidized, resulting in a very thick oxide film. Therefore, Tarcy [17] proposed some methods to prevent anode corrosion. For example, the content of Cu in Ni-Fe alloy should be increased without Ni, and the concentration of Al_2O_3 in electrolyte should be very high. The profile of the alloy ceramic anode in $NiFe_2O_4$/NiO/Cu/Ni after electrolysis for a long time at high Al_2O_3 concentration is shown in Fig. 3.18.

Fig. 3.17 Linear voltammetry of $NiFe_2O_4$/NiO-Ni cermet inert anode (solid line) and Pt anodes (dashed line). *x* voltage (V) (vs. Al/A^{3+}); *y* anodic current (mA)

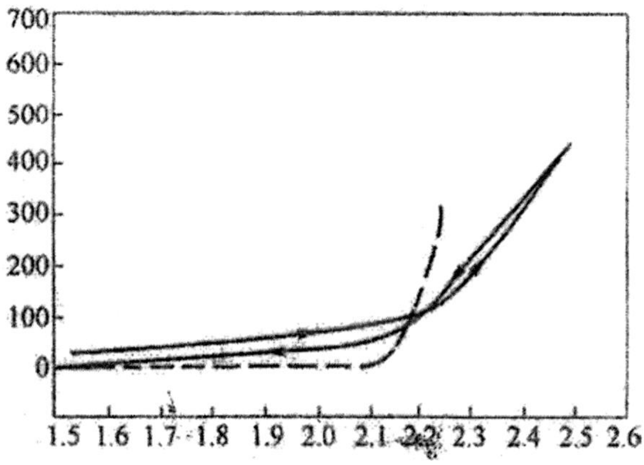

Table 3.10 Electrochemical corrosion test results of several metals and alloys

Type of metal or alloy	Corrosion current (A/cm)	Open-circuit potential (for Al)	Remarks
Ni	>1000	0.9 V	Serious corrosion
Pt	<10	1.9 V	No corrosion
Cu	123	1.87 V	Thick oxide film
Fe	140	1.34 V	Film, drop
Ta	400	0.9 V	Serious corrosion
Ni(40) + Nb(60)	>1000	0.9 V	Serious corrosion
Monel metal	100	1.42 V	Erosion pit
Inkang alloy	168	1.58 V	Production film
Inkang alloyx700	136	1.6 V	A thin film
Stainless steel (316, Low carbon)	500	1.1 V	A thin film
Hast alloy	80	1.3 V	Film
Carbo alloy	190	1.3 V	A thick film

Tarcy believed that Cu can be added to $NiFe_2O_4$/NiO ceramic matrix. These materials are readily available, cheap, and commercially feasible. Cu in ceramic anodes is protected by three factors. (1) Ceramics can be used as mechanical masking to avoid Cu falling off; (2) cathodic protection of copper-nickel alloy with high Cu at the beginning of electrolysis; and (3) copper aluminate ($CuAlO_2$) was produced by using electrolyte with high concentration of Al_2O_3 to stabilize Cu. In addition, the corrosion current, the open-circuit potential, and the depth of electrolyte permeating into the anode are the most important experimental parameters.

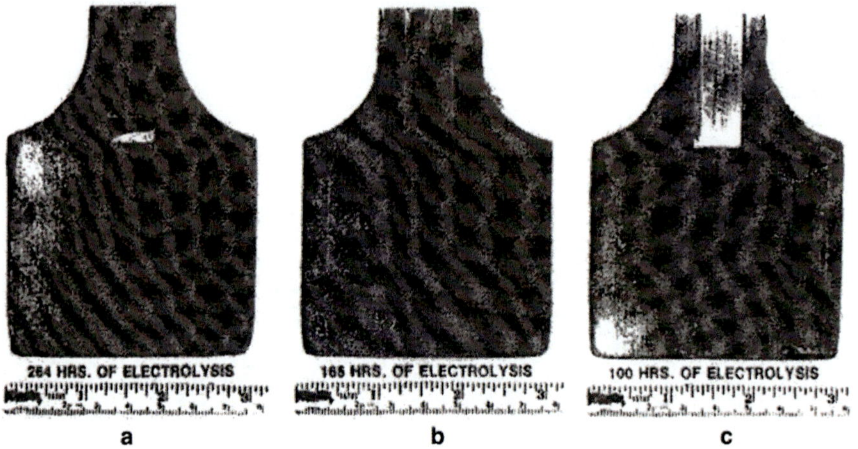

Fig. 3.18 The section diagram of $NiFe_2O_4/NiO/Cu/Ni$ cermet inert anode after long time electrolysis. (**a**) 24 h; (**b**) 165 h; (**c**) 100 h

3.6.3 Test of 2500 A Inert Anode Electrolyzer

Baker and Rolf (Alcoa Institute) [74] developed 2500 A inert anode electrolyzer. Ni-Fe-O series ceramic material was fabricated by isostatic pressing and welded to metal guide rod. After preheating at 1000 °C, the electrolysis is carried out in an electrolytic cell containing molten electrolytes. They believed that high Al_2O_3 concentration can reduce anode corrosion and the laboratory test can be used for future industrial trials. The self-heating test cell at 2500 A (Alcoa Institute) is shown in Fig. 3.19 [74].

The carbon anode was used when the electrolyzer started, and a stable production state is reached after a few days. Then, we moved it into the preheated inert anode. The electrolysis temperature is 960 °C, and the volume ratio of NaF/AlF_3 is 2.30. The content of Al_2O_3 reached saturation, so that no precipitation was formed. In this experiment, the inert anode used contains 60% NiO, 20% Fe_3O_4, and 20% Fe powder. After mixing, jet drying, isostatic pressing, and reactive sintering, a three-phase ceramic material was formed. It contains $NiFe_2O_4$ spinel, NiFeO oxide, and Ni/Fe alloy phase. The manufacturing process is shown in Fig. 3.20.

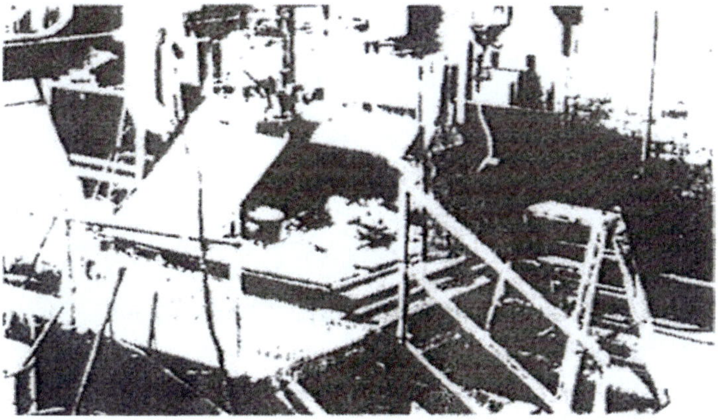

Fig. 3.19 2500 A electrolytic test cell for inert anode (Alcoa)

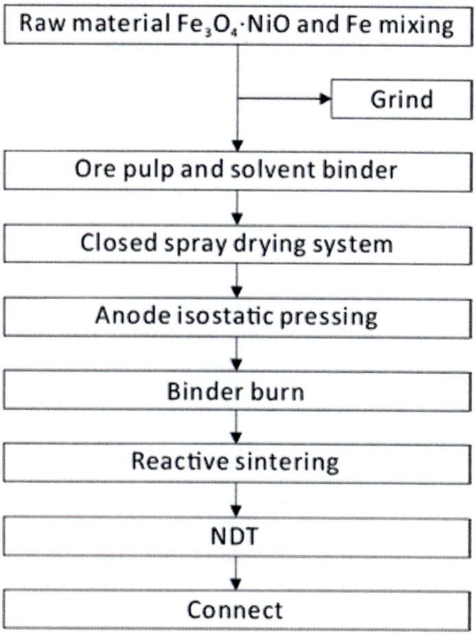

Fig. 3.20 Manufacturing process of Ni-Fe-based inert anode

The shape of the inert anode was originally flowerpot, with a nickel rod as the conductive connector and then a bottle shape. The current of the electrolysis cell was 2500 A. Figure 3.21 shows the profile of an inert anode after 21 days of testing.

The Fe and Ni contents in the aluminum decreased, and still greater than one with the electrolysis temperature 960 °C, as shown in Fig. 3.22.

Fig. 3.21 Photographs of inert anode cleaning after 21 days of test in 2500 A electrolytic cell

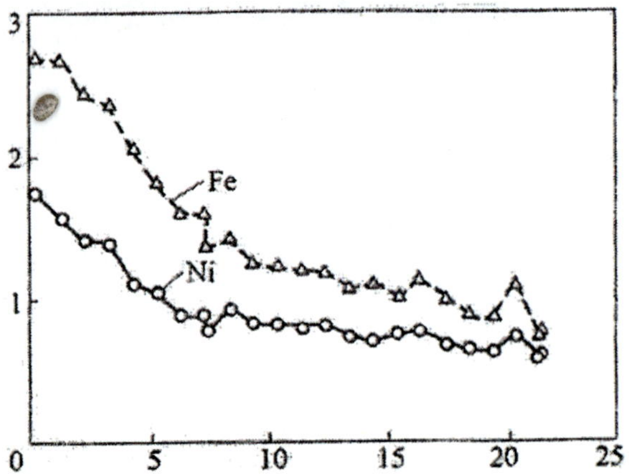

Fig. 3.22 Ni and Fe contents in aluminum after 21-day test in 2500 A electrolytic cell. *X* time (d), *Y* Fe or Ni%

3.6.4 *The Relationship Between Composition of the Inert Anode and the Corrosion Resistance of $NiFe_2O_4$-Based Cermet*

3.6.4.1 The Effect of Ceramic Phase Composition on Corrosion Resistance

The study of the corrosion resistance of cermet with different ceramic phases began in 1980. Alcoa began a series of studies on $NiFe_2O_4$-based cermet supported by the US Department of Energy [58]. They pointed out that 17Cu(Cu-Ni)-18NiO-$NiFe_2O_4$ cermet had good conductivity and corrosion resistance. Since then, 17Cu(Cu-Ni)-18NiO-$NiFe_2O_4$ cermet have gradually become the main research object of inert anodes. However, there is little evidence that the ratio of NiO to $NiFe_2O_4$ in ceramic phase is greater than 18% in determining the compositional fraction of NiO in ceramic phase.

DeYoung [58, 75] in 1986 pointed out that the activities of Ni and Fe components of $NiFe_2O_4$ in the aluminum electrolyte melt are inversely proportional to each other, satisfying Eq. 3.10:

$$k = 1/\left[\left(x_{Fe_2O_3}x_{NiO}\right)\left(y_{Fe_2O_3}y_{NiO}\right)\right] \tag{3.10}$$

The saturated solubility of NiO in cryolite melt is much lower than that of Fe_2O_3 (as shown in Table 3.7, 0.009% and 0.058%, respectively). It is suggested that the content of NiO in the ceramic phase of $NiFe_2O_4$-based cermet should be appropriately high. High content of NiO can reduce not only the content of impurity Fe but also the corrosion rate of anode and the total impurity content in primary aluminum.

In 1996, Olsen et al. [76] studied the electrolysis corrosion behavior of NiO in the ceramic phase with an excess of 0.17% or 23% in the ceramic phase. The results failed to effectively determine which ceramic phase of anode had the best corrosion resistance.

In a study published by Alcoa [66] in 2001, it was suggested that the formation of $NiAl_2O_4$ by NiO at high Al_2O_3 concentration can significantly reduce its solubility and protects the anode. However, Fe_2O_3 does not have the same case. This may also be one of the reasons for the NiO overdose. In addition, electrolytic measurements of a series of different anodized anodes are carried out in this report. Some impurities in the aluminum produced by anodic electrolysis are shown in Table 3.11. The high content of NiO reduces the impurity Cu, Ni, and Ag in aluminum, and the content of Fe did not decrease, as shown in Table 3.11.

Therefore, it is generally agreed that the appropriate excess of NiO can reduce the corrosion rate of Fe in cermet in the selection of ceramic phase. Then, it reduces the corrosion rate of the material and the total impurity content in primary aluminum.

3.6.4.2 Effect of Metal Phase Composition on Corrosion Resistance

Adding metal is beneficial to improving the sintering properties, mechanical properties, and electrical conductivity of cermet. However, the metal phase in the cermet is preferentially etched with respect to the ceramic phase. Therefore, the type and content of metal phase play a key role in whether cermet can become a qualified inert anode material.

In 1986, Alcoa [58, 77] showed the linear potential scanning of the cermet with Ni and Cu metal phase and found that the residual current of the former is greater than that of the latter. With the discovery of elemental surface scanning of polarized anode, the corrosion of the metallic phase in Ni-$NiFe_2O_4$-NiO cermet is more severe, and the electrolytes infiltrate more. Using metal Ni and Cu as working electrodes, it was found that the anode of Ni was pitting and no oxide protective layer was formed. The black substance appeared on the surface of Cu anode, which was determined by XRD analysis as Cu_2O and CuA_1O_2. It is concluded that the anode with Cu as the metal phase is passivated under the condition of anodic polarization, and the metal Cu is oxidized to form Cu_2O and CuA_1O_2 and attached to the metal surface, thereby slowing down the further corrosion of the anode. However, Ni cannot form a similar substance and an effective protective layer for inert anode. Therefore, it is recommended to use Cu or Cu-Ni alloy rich in Cu as metal phase of cermet. Tarcy [77]

Table 3.11 Comparison of impurity contents of cermet inert anodes with different ceramic phases in electrolytic primary aluminum

		Impurity content in primary aluminum (mass fraction)/%			
Anode number	Anode composition	Fe	Cu	Ni	Ag
776705–2	3Ag + 14Cu + 83$NiFe_2O_4$	0.375	0.13	0.1	0.015
776673–2	3Ag + 14Cu + 83"5324"	0.49	0.05	0.085	0.009

Note: "5324" refers to the $NiFe_2O_4$-NiO ceramic phase obtained after calcined with the mass ratio of NiO to Fe of 51.7:48.3

also pointed out that the cermet has no significance difference in corrosion resistance with metal phase contents of 5%, 10%, and 20%, respectively.

In 1987, Windisch [78] used cyclic voltammetry to study the electrochemical behavior of metal Cu, Ni, and cermet inert anode with Cu as metal phase during electrolysis. The corresponding current of Cu and cermet with Cu as metal phase is smaller, when the potential is lower than the decomposition voltage of aluminum oxide. The "residual current" is relatively small and close to that of the Pt electrode. The "residual current" corresponding to Ni is significant, indicating that there is obvious reaction. This result supports the conclusion of Tarcy [77] from the electrochemical point of view.

However, a large number of studies did not support that Cu or Cu alloys should be used as cermet metallic phases. Some researchers believed that the wettability of Cu and $NiFe_2O_4$ is not good, and the relative density of cermet is only 70%–80% under the condition that the deposit phase does not overflow and is evenly distributed. The temperature can effectively increase the density. However, the metal overflow and distribution are inhomogeneous, as shown in Fig. 3.23.

With Ni as the metal phase, the sintered properties of gold ceramics are good. The materials with 95% high density can be prepared under the condition that the deposit phase does not overflow and distributes evenly. The densification has a great influence on the corrosion resistance of the materials. Therefore, from the perspective of sintering densification, Cu is not conducive to the densification of cermet and the improvement of corrosion resistance.

Olsen [79] studied the electrolytic and metal Al migration of $NiFe_2O_4$-NiO-Cu anodic components under electrolytic conditions. The migration rate of element Ni and metal Al is about 50% of that of Cu and Fe. For NiO and other Ni compounds, Ni-based materials are more likely to be inert anode materials.

The anodic XRD analysis after electrolysis does not show the oxides of Cu and other passivating layer compounds as described by Tarcy. Therefore, the so-called anodic passivation did not occur in the corrosion of metal Cu.

In 2002, Lorentsen [80] carried out electrolysis experiment with 17Cu-$NiFe_2O_4$. The XRMA analysis of the anode surface after electrolysis showed that the Cu in the metal phase may migrate to the anode surface. There is no preferential corrosion of Ni in the Ni-rich metal phase as described by Tarcy. The migration rate of Cu in Cu-Ni alloy is two to three orders of magnitude higher than that of Ni. It is speculated that the migration rate of Cu in Cu-Ni alloy may be CuF or CuF_2.

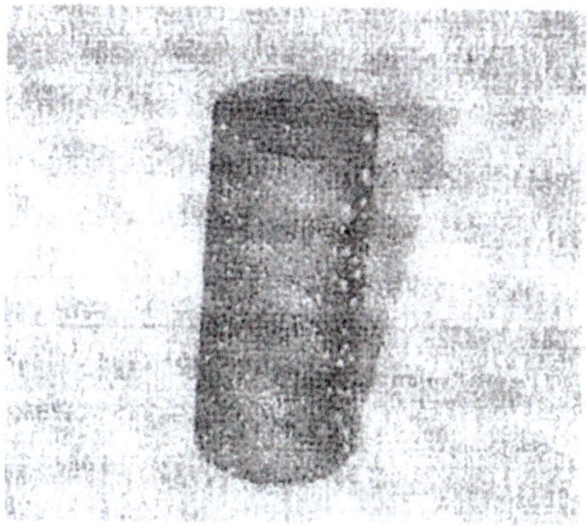

Fig. 3.23 Metal phase overflow during inert anode sintering of $NiFe_2O_4$-Cu cermet

In view of the above controversy about the influence of metal phase type and content on the corrosion resistance of $NiFe_2O_4$-based cermet, Li Xin Zheng [81] studied the effect of metal phase type and content on the corrosion resistance of M/(10NiO-$NiFe_2O_4$) cermet to Na_3AlF_6-Al_2O_3 molten salt. The microstructure of M/(10NiO-$NiFe_2O_4$) cermet inert anode corroded under the same electrolytic condition is different (Fig. 3.24). The surface structure of Cu/(10NiO-$NiFe_2O_4$) cermet is intact after electrolysis. The corrosion of materials is mainly caused by chemical dissolution corrosion of each component. It shows higher corrosion resistance than that of cermet with Ni and 85Cu15Ni as the main metal phase of electrochemical corrosion. Considering the influence of metal phase type and content on sintering properties, electrical conductivity and corrosion resistance of materials, metal Cu is suitable for 10NiO-$NiFe_2O_4$-based cermet inert anodes, and the content is 5%.

3.6.5 *Corrosion Mechanism of $NiFe_2O_4$-Based Cermet Inert Anodes*

With the development of research on corrosion resistance of $NiFe_2O_4$-based cermet inert anodes, the corrosion mechanism has been preliminarily understood, which can be classified into two categories: chemical corrosion and electrochemical corrosion [59, 82, 83]. Chemical corrosion can be classified into chemical dissolution, aluminothermic reduction, intergranular corrosion, electrolyte infiltration, etc. Electrochemical corrosion can be classified into anodic dissolution of the metal phase and electrochemical decomposition of the ceramic phase.

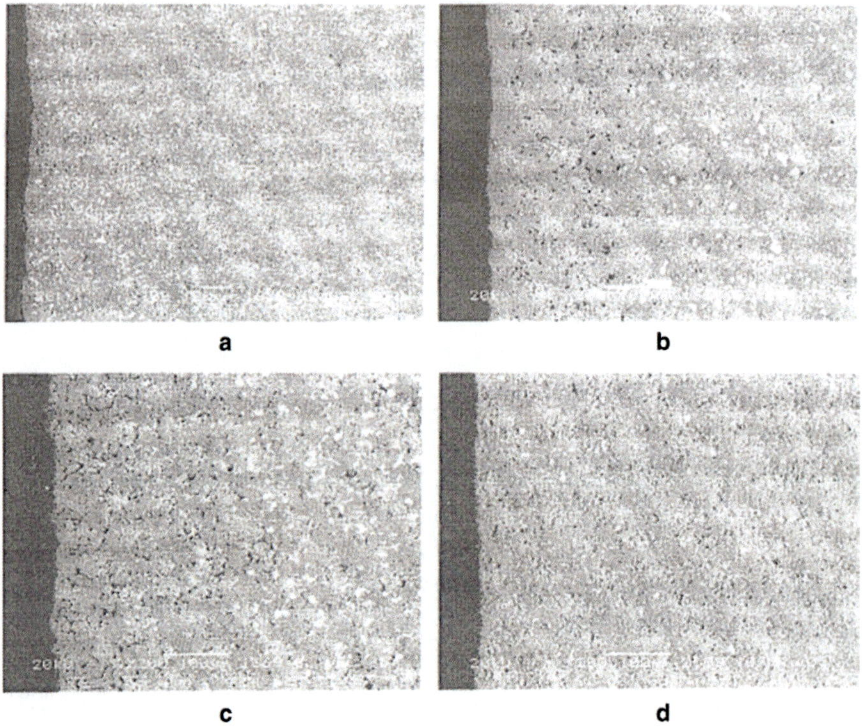

Fig. 3.24 SEM photos of M/(10NiO-$NiFe_2O_4$) cermet anodes after electrolysis. (**a**) 17Cu/(10NiO-$NiFe_2O_4$); (**b**) 17Ni/(10NiO-$NiFe_2O_4$); (**c**) 17(85Cul5Ni)/(10NiO-$NiFe_2O_4$); (**d**) 5(85Cul5Ni)/(10NiO-$NiFe_2O_4$)

3.6.5.1 Chemical Corrosion

Chemical Dissolution

Under the condition of aluminum electrolysis, $NiFe_2O_4$ ceramics are dissociated to a certain extent, and the dissociated NiO and Fe_2O_3 may be corroded by the reaction of Formulas (3.11) and (3.12):

$$3NiO(s) + 2AlF_3 \Leftrightarrow 3NiF_2(diss) + Al_2O_3 \quad \Delta G^{\ominus}_{1238K} = 75.51\,kJ/mol \quad (3.11)$$

$$Fe_2O_3(s) + 2AlF_3 \Leftrightarrow 2FeF_3(diss) + Al_2O_3 \quad \Delta G^{\ominus}_{1238K} = 179.20\,kJ/mol \quad (3.12)$$

The dissolved NiF_2 and FeF_3 may be reduced by Al dissolved in the electrolyte, as shown in Formulas (3.13) and (3.14), or migrated to the surface of the cathode aluminum solution and reduced into the aluminum solution by aluminum metal. Therefore, this promotes the chemical dissolution of NiO and Fe_2O_3 as well as the decomposition of $NiFe_2O_4$:

$$2Al + 3NiF_2(diss) = 3Ni + 2AlF_3 \quad \Delta G^{\ominus}_{1238K} = -947.83\,kJ/mol \tag{3.13}$$

$$Al + FeF_3(diss) = Fe + AlF_3 \quad \Delta G^{\ominus}_{1238K} = -481.73\,kJ/mol \tag{3.14}$$

The solubility of Fe_2O_3 in Na_3AlF_6–Al_2O_3 melts was studied by Diep [84]. It is concluded that with the reaction between Fe_2O_3 and AlF_3, NaF occurs in the form of Formula (3.15):

$$\frac{1}{2}Fe_2O_3 + \frac{1}{3}AlF_3 + xNaF = Na_xFeOF_{(1+x)} + \frac{1}{6}Al_2O_3 \tag{3.15}$$

Further determination shows that the solubility of Fe_2O_3 reaches the maximum value when the molar ratio is 3.0.

Aluminothermic Reduction

When there is a certain amount of aluminum solution in the electrolytic cell, the corrosion rate of anode is higher than that of electrolytic cell without aluminum solution [58]. This indicates that aluminum dissolved or suspended in the electrolyte is an important cause of anodic corrosion. The anodic corrosion reaction is shown in Formula (3.16):

$$2Al(s) + Fe_2O_3(s) = Al_2O_3(s) + 2Fe(s) \quad \Delta G^{\ominus}_{1238K} = -784.26\,kJ/mol \tag{3.16}$$

The thermodynamic calculation of Formula (3.16) shows that the reaction of metal oxides in the metal aluminum reduction anode has a considerable tendency.

In the study of corrosion resistance of inert anodes, the corrosion rate of electrified polarization is different from that of non-polarized electrolyte in the same electrolytes containing aluminum. The lower corrosion rate of the former is obviously due to the oxidation of the surrounding aluminum by the oxygen produced by the anode, which slows down the rate of thermal reaction. However, the current value should be controlled. If the current density is too low, it is not enough to restrain the reduction of aluminum.

If the current density is too high, the electrochemical corrosion of the anodic component and the wear corrosion (caused by the erosion of the anode gas) will be aggravated by the anodic polarization.

Intergranular Corrosion and Electrolyte Infiltration

The corrosion resistance of inert anode was studied by Wang Huazhang and Thostad [85]. The corrosion rate of inert anode is very high. The SEM analysis of the anode cross section after electrolysis shows that the electrolyte has entered into the internal pores of the anodes. In addition, in the microscopic grain gap, this can result in "intergranular corrosion" and the electrode swelling, spalling, and finally disintegrating. In addition, the electrode particles near the surface layer are infiltrated by the electrolyte as the electrolysis process goes on. When the metal phase is corroded, the ceramic particles are separated and isolated by the electrolyte, and even separated from the anode body into the electrolyte, resulting in increased corrosion.

3.6.5.2 Electrochemical Corrosion

Since the 1980s, researchers have tried to study the corrosion behavior of inert anodes under polarization through various electrochemical methods. Although some interesting conclusions have been obtained, most of these conclusions are scattered. The understanding of the electrochemical corrosion process of inert anode has not yet formed a unified understanding.

In 1986, Tarcy [17] used linear scanning voltammetry to study the corrosion of cermet, metal Ni, Cu, and inert metal Pt anode with different metal phases (Cu, Ni, 10Ni-90Cu) in electrolysis. There is a residual current in cermet anode relative to ceramic oxide and Pt anode. The residual current of Ni anode is greater than that of Cu anode, and the microzone of anode after electrolysis is synthetically analyzed. It is concluded that the corrosion resistance of cermet with Cu as the metal phase is better than that of the cermet with Ni as metal phase.

In 1987, Windisch [21, 86] used cyclic voltammetry to study the characteristics of voltammetric curves of metal Cu anode during electrolysis. The reduction peaks of oxidation were analyzed and may exist during the corrosion process. Cu was oxidized to Cu_2O and CuO. Cu_2O and CuO reacted with $A1_2O_3$ to form $CuAlO_2$. The voltammetric curves of cermet anode with Cu as metal phase were studied.

In 2001, Lorentsen [87] studied 17Cu-$NiFe_2O_4$ by linear scanning and AC impedance method and measured the relative parameters (such as external impedance, electrolyte resistance, n value of CPE element, slope of Tafel curve, etc.). In a report submitted by Northwest Aluminum Technology Corporation in 2003 [88], the laboratory phase of the "impedance spectrum" study was completed in December 2002. The "impedance spectroscopy" and "resistance measurement" on the test tank were completed in December 2001. However, the report did not disclose further details.

Electrochemical corrosion process of the inert cermet anode generally includes the anodic dissolution or decomposition of the metallic and the ceramic phases. The metal phase in the ceramic anode is added to improve the electrical conductivity of ceramic matrix. However, it has relatively strong electrochemical activity under the condition of anode polarization. The anodic discharge of the oxygen complex ion in the melt and the release of oxygen not only occur on the anode but also lead to anodic dissolution of the metal phase. In addition, the corresponding complex ion is formed in the melt. This leads to the consumption of the anode [58, 77]. Taking Ni as an example, when anodic dissolution occurs, the electrolytic reaction can be expressed as Formula (3.17):

$$3\text{Ni}(\text{s}) + \text{AlF}_3(\text{s}) \Leftrightarrow 3\text{NiF}_2(\text{s}) + 2\text{Al}(1) \tag{3.17}$$

At 1238 K of Formula (3.17), E_{1238K} = 1–637 V. This indicates that such reactions may occur under the lower voltage of the normal Al_2O_3 decomposition reaction of Formula (3.2) and lead to "residual currents" in electrochemical measurements. The electrochemical corrosion of the ceramic phase is the anodic decomposition of the ceramic phase under anodic polarization condition. Oxygen element is oxidized at the anode to produce oxygen. The corresponding metal element acts with melt to form complex ion and enter melt, resulting in anode consumption. Taking Fe_2O_3 as an example, when electrochemical corrosion occurs, the electrolysis reaction can be expressed as Formula (3.18):

$$\text{Fe}_2\text{O}_3(\text{s}) + 2\text{AlF}_3(\text{s}) = 2\text{FeF}_3(\text{s}) + \frac{3}{2}\text{O}_2 + 2\text{Al}(1) \tag{3.18}$$

At 1238 K, E_{1238K} = 2.51 V. Although it is higher than the decomposition voltage of Al_2O_3 (Formula 3.2), the reaction may also take place under the conditions of higher anode polarization potential and lower Al_2O_3 concentration.

3.6.6 Preparation of Inert Anode for $NiFe_2O_4$-Based Cermet

The preparation process of the cermet inert anode plays a decisive role in its corrosion resistance, electrical conductivity, mechanical properties, and so on. The relationship between the preparation process and the properties can be characterized by the microstructure of the material, including the type, quantity, and structure of the phase. The properties of the materials will be greatly changed by adopting different technological methods to change the microstructure.

Since 1980, the preparation of $NiFe_2O_4$-based cermet inert anodes was based on the traditional powder metallurgy technology. The representative is Alcoa process [58]. Its inert anode research group has studied the preparation technology of large cermet electrode for 3 years, and successfully prepared large anode with 63 mm, and investigated it in the electrolytic cell of 2500 A. The proposed technical route for the preparation of materials is shown in Fig. 3.25. This includes the selection of oxide raw materials (average particle size of l μm), blending, calcination, spray drying, adding metal powder by ball milling and blending (Ni, Cu powder with average particle size of 10 μm), spray granulation, isostatic pressing, wet blank processing, entering (controlling oxygen content in atmosphere), etc. In the study, they also used hot pressing sintering process to prepare cermet. However, because of the reaction of oxide raw materials with graphite mold, the high cost, and the difficulty of preparing large size special-shaped materials, it was not used in future research.

The preparation process of $NiFe_2O_4$-based cermet anode was investigated by Olsen [89]. With cold isostatic pressing (CIP) (pressure of raw material powder was 300 Mpa), the blank density reached 60% of the theoretical density. The sintering

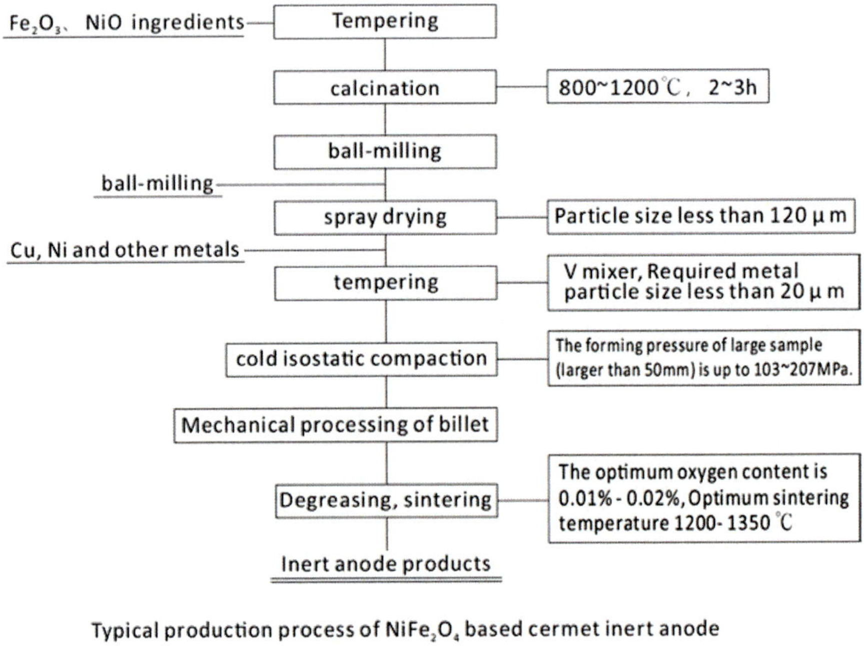

Fig. 3.25 Typical production process of $NiFe_2O_4$-based cermet inert anode

process was the same as that reported by Weyand [58], and the maximum sintering temperature under the protection of argon is 1350 K. Sometimes the metal phase overflow phenomenon is observed. The samples were analyzed by SEM and XRD, and no other impurity phases were found.

The preparation technology of cermet inert anode has been systematically studied in Central South University of China [58, 66, 74, 79, 88–91]. The cylindrical samples of φ 20 mm × 40 mm, φ 50 mm × 60 mm and the deep cup samples of φ 110 mm × 130 mm (diameter × height) were successfully prepared. The results show that $NiFe_2O_4$-based ceramic powder should be synthesized in air. The oxygen partial pressure should be controlled when sintering $NiFe_2O_4$-based cermet. Cu is easy to overflow. M-$NiFe_2O_4$ cermet is easier to realize densification than Cu-$NiFe_2O_4$ and Cu-Ni-$NiFe_2O_4$ cermet. Ni is a good sintering aid, etc.

MFe_2O_4-Ni-Cu-NiO cermet were prepared through hot pressing and sintering process at Northeast University and Tsinghua University of China [92]. They found that increasing the temperature is beneficial to increasing the density, and 1000 °C is the upper limit. If the temperature is high, the density will decrease. After 6 h of electrolysis at a current density of 1.0 A/cm^2, the anode surface is angular without obvious corrosion trace.

To sum up, a lot of research has been done on the preparation technology of $NiFe_2O_4$-based cermet. However, sintering temperature, sintering atmosphere, and the behavior of metals in the sintering process need further study to improve the comprehensive properties of anode in the treatment of raw materials.

3.6.7 Sintering Densification of $NiFe_2O_4$ Based Cermet Inert Anode

The inert anode material used in aluminum electrolysis should have not only the target phase composition to ensure good electrical conductivity, thermal shock resistance, and corrosion resistance but also high density to resist the infiltration of high-temperature aluminum electrolyte melt [9]. The increase of material density can improve the electrical conductivity, enhance the mechanical strength, avoid the oxidation of the metal phase, etc. Sintering is a process in which the total surface energy and defect concentration are reduced and densified under the action of high-temperature thermal energy. In order to obtain a high-density cermet material with a target phase composition, the operating conditions of the sintering atmosphere, sintering temperature, and holding time should be strictly controlled. The properties of Mg^{2+}-doped $NiFe_2O_4$ inert anode materials were studied by Berchmans [90]. The blank was sintered at 1000 °C for 50 h in air atmosphere. Mg^{2+} embedded in the lattice of $NiFe_2O_4$ spinel will increase the cell parameters and form a new substance $Ni_{0.4}Mg_{0.6}Fe_2O_4$. Mg^{2+} preferentially occupies B site (substitution Ni^{2+}) of spinel cubic structure and partially occupies A position (Fe^{3+}). The sintering properties, electrical conductivity, thermal shock resistance, and corrosion resistance of the materials were improved.

Yao Guangchun et al. [93] studied the effect of MnO_2 doping on the sintering process and microstructure of Ni-Fe spinel inert anode materials. Adding 1% MnO_2 can refine the grain size, and the particle size distribution is uniform. This promotes the sintering and improves the bending strength.

3.6.8 Mechanical Properties of $NiFe_2O_4$-Based Cermet Inert Anode

When inert anode is applied to aluminum electrolysis, it must be preheated and started. The purpose of the inert anode preheating is to approach or reach the normal production temperature of the electrolytic cell through slow heating for a certain time. This can avoid the "thermal shock" during start-up and lead to the electrode cracking. Although the preheating and starting process is very short in the whole life of inert anode, it has a decisive effect on the life of inert anode. Therefore, the mechanical properties and thermal shock resistance of inert anode are very important for the industrial application of inert anode.

The mechanical properties of inert anodes should not be considered in the test of small sample and low current. However, in large-scale test, the mechanical properties of oxide and cermet inert anode are very important. In 1991, Reynolds Metals Company carried out a 25-day continuous electrolysis test with cermet anode in a 6000 A electrolytic cell. The main problems of exposure are poor thermal shock resistance of large-scale anodes, electrode cracks, serious damage of conducting rod, and the final failure of the test, as shown in Fig. 3.15.

In the Inert Anode Report of Alcoa in 1986 [58], four points of bending strength, Wechsler's modulus, fracture toughness, Young's modulus, shear modulus, and Poisson's ratio were tested for 51.1% NiO-48.3% Fe_2O_3 (i.e., 5423), 20% Fe-60% NiO–20% Fe_2O_3 (i.e., 6846), 5423+:30% Ni, 5423 + 17% Cu, and other materials. The results are shown in Table 3.12, where A represents $NiFe_2O_4$. Alcoa announced in a report published in July 2001 that the application of inert anodes will be delayed

Table 3.12 Performance of inert cermet anode reported by Alcoa in 1986 [58]

Performance index	Al.17%u	Al.30%Ni_b	Al.30%Ni_a	Al-66	6846	6846	5324
Preparation method	Calcine sintering	Calcined sintering	Calcined sintering	Calcined sintering	Calcined sintering	Reactive sintering	Calcined sintering
Theoretical density/g cm volume	6.28	6.55	6.5 5	6.35	6.35	6.35	5.72
Density/g cm	6.0	6.52	6.55	6.11	6.12	5.89	5.69
Open porosity %	0.06	0.09	0.11	0.2	0.3	1.94	0.16
Four-point bending strength/MPa	104	182.9	192.4	126.8	112.8	105.8	165.6
Wechsler's modulus/GPa	12.4	20.9	7.8	4.9	15.4	8.0	13.1
Fracture toughness/ MPa $m^{1/2}$	–	5.15	5.43	3.64	3.75	4.84	1.92
Young modulus/GPa	145	–	–	175	–	146	155
Shear modulus/ GPa	55.8	–	–	–	–	56	63
Poisson ratio	0.3	–	–	–	–	0.29	–
Microstructure metal content %	20	39	40	31	30	28	3
Microstructure (cavitation rate) %	1.2	1.5	1.5	1.6	4	5.3	3.4
Phase	A, NiO	A, NiO	$NiFe_2O_4$	A, NiO	A, NiO	A, NiO	A, NiO
Composition	Ni, cu	Ni	NiO, Ni	Ni	Ni	Ni	

Note: A for $NiFe_2O_4$; Al stands for "5324"

again due to thermal shock cracking and conductive connection failure of inert anodes [73]. The mechanical properties of $NiFe_2O_4$-based cermet inert anode play an important role in its large-scale test. However, there are relatively few studies on the mechanical properties of $NiFe_2O_4$-based cermet inert anode.

3.6.9 High-Temperature Oxidation Resistance and Electrical Conductivity of $NiFe_2O_4$-Based Cermet

Aluminum electrolytic inert anode should have good high-temperature stability and high electrical conductivity. The change with temperature should not be too huge at high temperature. Otherwise, the distribution of anode current density will be uneven, and sometimes the connection between the anode and the metal guide rod will be broken due to excessive current concentration. In the inert anode of aluminum electrolysis during the electrolysis process, a large amount of O_2 is precipitated on the surface of the electrode, which will oxidize the metal phase in the surface area of $NiFe_2O_4$ cermet. This changes the composition of the material phase and affects the conductivity of the material, etc.

The conductivity of $NiFe_2O_4$ ceramics at 950 °C is about 2 s/cm. Compared with the current conductivity of carbon anode in the aluminum electrolysis industry (the conductivity is about 200 s/cm under electrolytic condition), the conductivity of $NiFe_2O_4$ ceramics is significantly lower. Even at high temperatures, the requirements for inert anode materials cannot be met. After adding metal Fe, there will be Ni-Fe phase in the material, which can improve the conductivity of the material. The conductivity of the material can reach 70 s/cm at high temperature. The electrical conductivity of $NiFe_2O_4$-based cermet with 17% Cu is about 90 s/cm at 1000 °C [58]. Alcoa [94] improved the conductivity of cermet inert anodes by adding metal Ag. Lai [95] established a high-temperature conductivity measurement device in response to the reported difference in high-temperature conductivity of inert anodes. As the temperature changes, the electrical conductivity of $NiFe_2O_4$-based ceramics of different types and metal phase contents were systematically measured. Cu-Ni alloy is more favorable to improve the electrical conductivity of $NiFe_2O_4$ ceramics. The conductivity of 20% Ni/$NiFe_2O_4$ cermet is 69.41 s/cm at 960 °C.

3.6.10 Connecting Technology of Inert Cermet Anode and Metal Guide rod

In the inert anode technology, the reliable connection between anode and metal guide rod is one of the key problems that needs to be solved in design and manufacture. It is also a difficult problem in this field. The main reason is that the environment of material use has put forward the strict requirements to the connection material. (1) Under the operating temperature of 900–1000 °C, the joint is required to have not only sufficient mechanical strength but also a long service life. (2) The joints are required to have good creep resistance, thermal fatigue, thermal shock, and corrosion resistance, so as to ensure the stable operation of the interface under the condition of high temperature and corrosive gases such as oxygen, fluoride, etc. (3) The bonding material should have good plasticity, yield strength, and tensile strength. Thermal expansion coefficients are close to the bonding material. This can ensure that the interface does not produce high residual stress during the temperature cycle. (4) Because of the low thermal conductivity, poor electrical conductivity, and weak thermal shock resistance of ceramics, it is required that the electrical conductivity of the connectors should be matched as much as possible to avoid the appearance of the residual thermal stress and the resistance thermal phenomenon at the interface. In fact, it is quite difficult to satisfy the above conditions at the same time. At present, there are three kinds of bonding methods between ceramics and metals: mechanical connection, welding connection, and chemical bonding, among which welding is divided into brazing and diffusion welding. The mechanical connection and diffusion welding methods are commonly used to prepare cermet inert anodes.

3.6.10.1 Mechanical Connection

Mechanical connection is a commonly used inert anode connection method in early research. It mainly includes bolt connection and brazing spring compression mechanism connection, as shown in Fig. 3.26 [58, 60, 96]. After electrolytic test of Alcoa with the inert anode in 2500 A electrolytic cell, the connection effect of brazing spring compression mechanism is better than that of threaded connection [58]. The special points of bolt connection are simple operation and detachability. The machining diagram of the inert anode sample connected by thread, the profile of the material object, and the sample after electrolysis are shown in Fig. 3.26. The test results show that the joints with threaded joints have poor electrical conductivity and poor connection stability. Moreover, due to the poor processing performance of cermet materials, the processing yield is low and the cost is high. The anodic machining area is the area where defects are most likely to cause defects during the electrolysis process and lead to anode failure, as shown in Fig. 3.26c. Strachan [96] used a brazed spring compression mechanism to connect the anode to the conductive connecting rod, as shown in Fig. 3.26d. The method first brazed the guide rod and the anode matrix and then strengthened the connection part with the spring mechanism. This can obtain a relatively stable connection. However, due to the thin brazing layer, the stable operating temperature of the solder cannot meet the requirements of aluminum electrolysis test. Therefore, the joint is always subjected to compressive stress, which is prone to stress concentration.

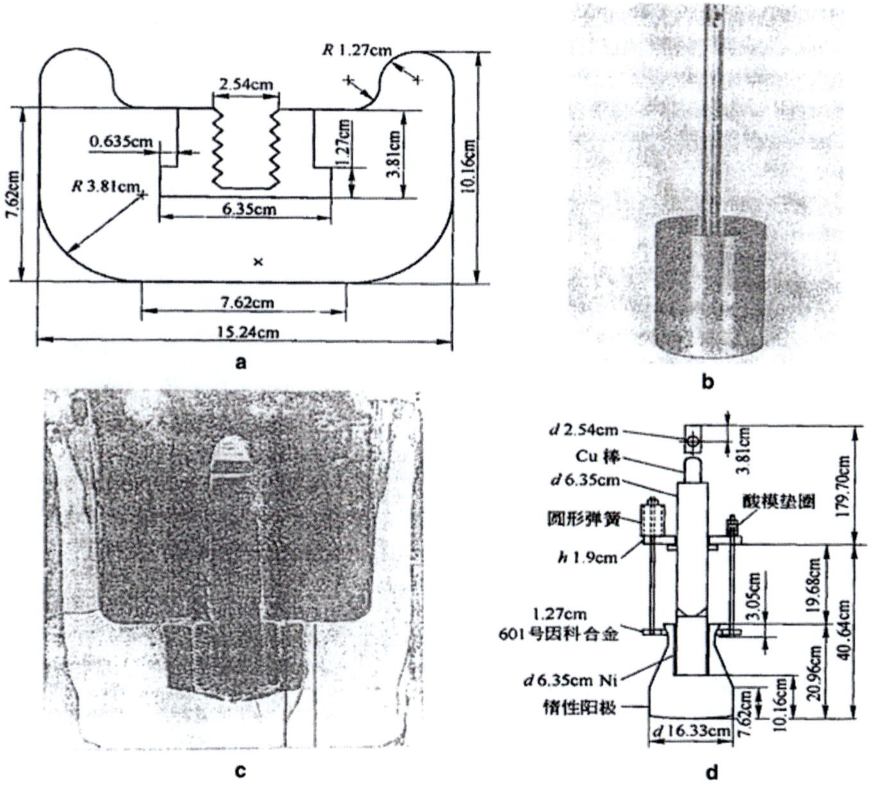

Fig. 3.26 The mechanical connection between inert anode and metal guide rod. (**a**) Thread connection anode machining sample [60]; (**b**) thread connection object [58]; (**c**) anode section of electrolytic back thread connection [58]; (**d**) brass solder + spring compression connection mechanism [58, 96]

3.6.10.2 Welding Connection

In order to realize the connection between high-temperature-resistant ceramics and metals, brazing, solid phase diffusion welding, and transient liquid phase diffusion bonding are mainly used at present. The brazing method of high-temperature resistance brazing has been more studied and has better connection properties. The main problems are that the brazing temperature is high, the suitable surface is narrow, and the high-temperature strength is not ideal. The characteristics of brazing joint are diffusion, physical force, chemical bond action, high strength, high air tightness, high-temperature resistance, and high reliability. However, the process is difficult, the cost is high, and the joint has internal stress.

In the field of diffusion welding of cermet inert anodes and conductive steel rods, Weyand [58] and Peterson [97] reported the test scheme and results of bonding of

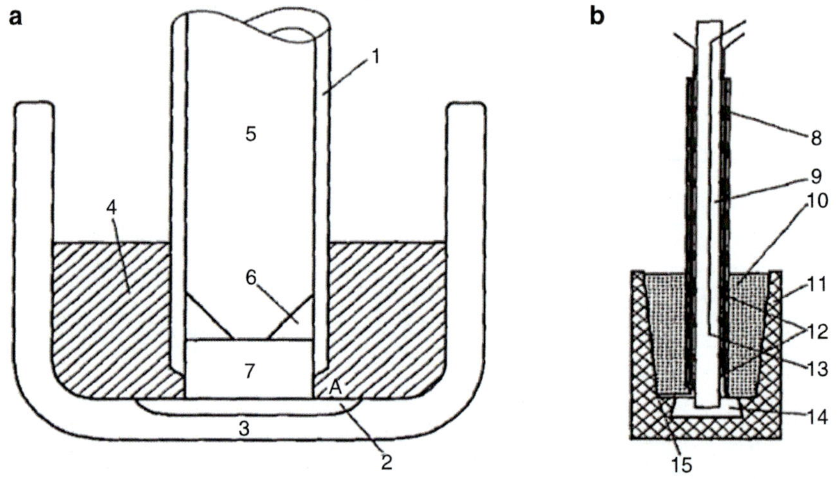

Fig. 3.27 Diffusion welding of inert anode and metal guide rod. (**a**) Using Ni connecting rod [52]; (**b**) using Cu-Ni alloy connecting rod [98]. (*1*) inconel alloy, (*2*) inert anode reduction zone, (*3*) cermet anode, (*4*) inert anode fragment filler, (*5*) carbon steel, (*6*) welding tape, (*7*) Ni_{201}, (*8*) corundum casing pipe, (*9*) anode guide rod, (*10*) Al_2O_3 filler, (*11*) inert anode, (*12*) guide rod potential detection point, (*13*) thermocouple, (*14*) Cu-Ni alloy, (*15*) anode potential detection point

cermet inert anodes and conducting rods by solid phase diffusion welding, respectively. Its structure is shown in Fig. 3.27.

Figure 3.27a is a model of diffusion welding joint of inert anode developed by Weyand et al. Ni_2O is used as the intermediate layer to connect cermet and anode steel rod. Before the diffusion welding, the surface of the anode joint should be metallized, and the surface of the material should be reduced by a reducing agent to obtain a metal layer. This can construct a gradient structure from cermet to Ni_2O. Under the influence of processing technology and material characteristics, it is easy to produce holes and inclusions in transition metal layer during metallization of cermet surface, which will become the main vacancy source in the material. The microstructure is very unstable when it works at high temperature for a long time. The movement of the vacancy at high temperature results in the decrease of the connection strength and even the failure of the connection. Figure 3.27b is a diagram of diffusion welding structure and single inert anode assembly using metal Cu-Ni connecting anode and anode steel rod developed by Peterson et al. [97].

A piece of Cu-Ni alloy was sintered at the anode junction, and then the anode rod was connected with Cu-Ni alloy and anode cup by diffusion bonding. The connection mode has the characteristics of stable connection and high connection strength. However, the process is too complicated. Alcoa [58] has investigated the structural stability of diffusion welding between cermet anode and Ni rod in the 60 A electrolytic cell. The short-time electrolysis test shows that the diffusion welding is very successful. However, the results of 2500 A electrolysis experiment show that the strength of the diffusion joint decreases during the process of operation and even causes the anode to fall off directly.

3.7 Low-Temperature Aluminum Electrolysis

Low-temperature aluminum electrolysis refers to the process of aluminum electrolysis at 800–900 °C or lower temperature. It is considered as the most potential energy-saving and consumption reduction technology. It is also the main way to solve the corrosion resistance problem of inert anode. It has become one of the most active topics in the international aluminum metallurgy field.

The melting point of aluminum is 660 °C, and the temperature of aluminum electrolysis can be controlled above 700 °C to meet the requirement of producing liquid aluminum through cathode. Therefore, when the production process of Hall-Heroult aluminum electrolysis was put forward, its inventor had envisaged low-temperature electrolysis. Low-temperature electrolysis can reduce the heat loss of the electrolytic cell and improve the current efficiency. This reduces the energy consumption and cost of primary aluminum production. However, the most fatal weakness of low-temperature electrolytes, namely, the difficulty of alumina dissolution (low dissolution rate and solubility), seriously hinders its development and application.

The development of any kind of inert anode (ceramic, alloy, or cermet) has encountered a common problem, that is, the corrosion resistance of inert anode (for ceramics and cermet, there is also thermal shock resistance) cannot meet the requirement of current aluminum electrolyte system and electrolytic process (characteristic of high-temperature and low-alumina concentration). To solve the corrosion resistance problem of inert anode, it is necessary to provide a better service environment for inert anode in addition to further improving the properties of materials. It is mainly a new electrolyte system with the characteristics of "low-temperature and high-alumina concentration" and a new electrolytic process. The reduction of electrolysis temperature can significantly reduce not only the oxidation rate of metal phase (or metal matrix) (the oxidation rate of metal can be reduced by one order of magnitude if the temperature is decreased by 100 °C) but also the dissolution rate of ceramic phase. These two aspects are the main causes of inert anode corrosion failure [99]. This demand has greatly promoted the study of low-temperature electrolytes.

The low-temperature electrolyte of inert anode can basically be divided into two systems: NaF-AlF_3 and KF-AlF_3. Both of them can decrease the melt primary crystal temperature by reducing the molar ratio of electrolyte or adding other additives. This can realize low-temperature electrolysis.

3.7.1 *NaF-AlF_3 Low-Temperature Electrolyte System*

In 1994, Beck [49] used Fe-Cu-Ni alloy anode to conduct electrolysis experiments in low-temperature electrolytes of NaF-AlF_3 (or adding part of KF and LiF) at 750 °C. The related electrolysis process was studied. The disturbance of the anode bubble can suspend the undissolved Al_2O in the electrolyte melt. This can effectively supplement the Al_2O consumed in the electrolysis process. In addition, the Al_2O consumed in the electrolysis process can be effectively replenished. The vertical multi-chamber electrolytic cell structure can make the space utilization ratio of the electrolytic cell more than 20 times that of the traditional electrolytic cell. In 1995, a further electrolytic test was carried out for 300 A, which reached the impurity Cu content in primary aluminum.

0.3% is in the initial stage of electrolytic cell start-up. The content of impurity Ni, Fe, and Cu was less than 0.5% and 0.03% (the quality requirements of primary aluminum have been met.) after 2 days [100]. Since then, a series of studies have been carried out on low-temperature electrolysis of alloy anode with low molar ratio NaF-AlF_3. A variety of Al_2O suspension electrolytes (also known as slurry electrolytic cells) have been proposed [98, 101–105]. From 1998 to 2004, the "study on anode life in low-temperature electrolysis" carried out by Northwest Aluminum Technology Corporation and other companies, supported by the US Department of Energy, represents the latest progress of studies [52].

Alcoa [64] conducted a long-term (200 h) electrolytic corrosion test in a low-temperature electrolyte of 36% NaF-60% AlF_3 (cr = 1.12), which was considered as an ideal low-temperature electrolyte by using the developed 5324–17 Cu cermet anode with the aid of the US Department of Energy. The electrolysis temperature is 800 °C, and the anode current density is 0.5 A/cm^2. During the electrolysis process, the melt is stirred by gas to accelerate the dissolution of alumina. The high purity alumina crucible was used for dissolution and consumption, and excessive Al_2O was added to maintain a high Al_2O concentration. The alumina content measured during electrolysis was 4.05%–4.5%. However, the test results of are not satisfactory (some experiments have obtained lower annual corrosion rate of 0.254–0.762 cm). The main problem is that the circulating electrolyte is favorable to the dissolution of Al_2O. However, the electrolyte melt contains metal Al, which directly corrodes the anode. Therefore, using O_2 instead of Ar as stirring gas can slow down the reduction of anode by aluminum. This results in the formation of a nonconductive layer on the anode surface. Thus, it is necessary to separate the dissolution zone of alumina from electrolysis to solve this problem.

3.7.2 KF-AlF_3 Low-Temperature Electrolyte System

Compared with the NaF-AlF_3 system, the KF-AlF_3 system has the advantages that the alumina dissolves faster and higher solubility. However, the permeation and expansion of K to carbon material is about ten times that of Na. This is fatal to the electrolytic cell with carbon material as cathode and lining. For this reason, there are few reports on low-temperature aluminum electrolysis in KF-AlF_3 system. In recent years, more and more people have realized that the successful development of inert anode must has a low-temperature aluminum electrolyte system, which can provide "low-temperature, high-alumina concentration" environment. However, it is difficult to obtain high-alumina concentration in NaF-AlF low-temperature electrolyte system. It is expected that new cathode materials (such as TiB_2 cathode) and liner materials (such as corundum) will appear in view of the percolation failure of carbon lining in KF-AlF_3 system. Therefore, low-temperature electrolysis of inert anode in KF-AlF_3 system has attracted more attentions in recent years.

In 2004, low-temperature electrolysis experiments were carried out with Cu-Al metal anode and TiB_2 cathode for 10 A, 20 A, and 100 A, respectively, in 50% AlF_3–45% KF-5% Al_2O_3 electrolyte at 700 °C by J. H. Yang [47]. The electrolysis process lasted up to 100 h, the anodic current density was 0.45 A/cm^2, the current efficiency was 85%, and impurity content of Cu was less than 0.2%, as shown in Table 3.13. Therefore, Cu-Al alloy anode is expected to be successfully applied in KF-AlF_3-Al_2O_3 electrolyte system, and the further research is necessary.

J. H. Yang [48] used Al-Cu alloy anode and TiB_2 wettable cathode to carry out a series of 100 A-100 h electrolysis experiments in KF-AlF_3 low-temperature electrolyte melt. The effects of NaF content, current density, and electrolysis temperature on the corrosion resistance of the anode were studied. It is considered that the requirements for conducting a larger-scale electrolytic test have been met. In 2007, J. H. Yang studied the solubility of Al_2O_3 in KF-AlF_3 electrolyte melt.

In Moltech's inert anode electrolysis test, in order to enhance the solubility of alumina in electrolyte melt, electrolytes containing KF were used, such as Na_3AlF_6 + 11% AlF_3 + 4% CaF_2 + (5%–7%) KF + (7%–8%) Al_2O_3 [54], Na_3AlF_6 + (10%–14%) AlF_3 + (2%–6%) CaF, + (3%–7%) Al_2O_3 + (0%–8%) KF [57] or Na_3AlF_6–11% AlF_3–4% CaF_2–7% KF-9% Al_2O_3 [55].

J. W. Wang [106] carried out 5% Cu, 9.5% NiO, and 85.5% $NiFe_2O_4$ anode study on electrolytic corrosion in different low-temperature electrolytes, including K_3AlF_6-Na_3AlF_6-AlF_3 + 5% Al_2O_3 and 50% AlF_3 + 45% KF +5% Al_2O_3.

In recent years, there have been reports on KF-AlF_3 low-temperature electrolyte system in Russia. In Kryukovsky [107], to solve the problem that the conductivity of KF-AlF_3 electrolyte system becomes worse after lowering the temperature, KF-AlF_3-Al_2O_3 (CR = 1.3), KF-AlF_3-yF (CR = 1.3), and KF-AlF_3-Al_2O_3-LiF (CR = 1.3) have been studied about the change of electrolyte melt conductivity with temperature, Al_2O_3 content (0–4.8%), and LiF content (0–10%) at 680–770 °C. The conductivity of KF-AlF_3 low-temperature electrolytes is lower than that of existing electrolytes. However, the addition of LiF has improved this condition. It is considered that the low molar ratio KF-AlF_3 melts added with LiF can be used as low-temperature electrolytes in new electrolytes.

3.7.3 *Main Problems That Need to Be Solved in Low-Temperature Aluminum Electrolysis of Inert Anode*

3.7.3.1 Alumina Dissolution

The difficulty of dissolving Al_2O_3 at low temperature is the main problem of low-temperature electrolysis and the biggest obstacle to the application of inert anode in low-temperature electrolyte system of NaF-AlF_3. The solubility of alumina

Table 3.13 Impurity content in aluminum metal under different electrode conditions

Experiment number	Current/A	Time/h	ω(Cu)/%	ω(Fe)/%	ω(Si)/%	ω(Ni)/%	ω(Cr)/%	ω(Mo)/%	ω(K)/%	ω(Na)/%	ω(Mn)/%
AlT22	10	31	0.51	0.032	<0.01						
AlT25	10	100	0.51	0.0359		0.0276			0.0006	0.002	0.0051
AlT53	20	32.5	0.1	0.24	<0.01	0.03	0.05		<0.025		
AlT55	20	56.4	0.16	0.19	<0.01	0.02	0.04		<0.025		
AlT57	100	50	0.09	0.03	<0.01	<0.01		0.03	0.021	<0.01	

decreases significantly (from 10% to 3% in NaF-AlF_3 system), and the dissolution rate significantly decreases. Although suspension electrolysis can increase the concentration of Al_2O_3 in melt, it will cause new engineering technical problems, especially the gas disturbance is not conducive to the accumulation of metal Al. It will not only affect the efficient operation of electrolytic cell but also aggravate the corrosion of anode. Some researchers have proposed to adopt active Al_2O_3 with high surface area, but its water absorption is very strong. It is difficult to meet the requirements of transportation, storage, and discharge in industry. According to the present results, Al_2O_3 quickly dissolves in KF-AlF_3 low-temperature electrolyte system, which may solve this problem. In addition, it is necessary to develop the Al_2O_3 blanking technology in the new electrolyte system to effectively disperse Al_2O_3 in the electrolyte melt and promote the dissolution.

3.7.3.2 Electrolyte Cathodic Shell

Low-temperature electrolysis is mainly achieved by reducing the molar ratio of electrolyte system. From the NaF-AlF_3 and KF-AlF_3 phase diagrams, the liquid-phase line of the system becomes steeper in the low molar ratio range (i.e., the initial crystal temperature is more sensitive to the molar ratio) [108]. During electrolysis, Na^+ or K^+ electromigration and cathodic reduction of aluminum complex ions lead to the enrichment of Na^+ or K^+ in the cathode region, which leads to the increase of melt molar ratio, the increase of initial crystal temperature, and the decrease of superheat in cathode region. When the conductivity becomes worse, solid electrolyte precipitates in the cathode and forms a layer of hard shell, that is, the cathode shell. The cathode shell can make the cathode conductivity become worse, the cell voltage increases, and the electrolysis process can be blocked seriously. It is generally considered that the solution to this problem is to increase the superheat, which is one of the bases for determining the optimum electrolysis temperature for low-temperature electrolysis. In addition, the cathodic crust can be reduced by controlling the cathodic current density properly.

3.7.3.3 New Cathode and Lining Material

Under the condition of low-temperature electrolysis, it may be necessary to maintain a higher superheat degree than the current electrolyte. It is difficult for the side part of the electrolytic cell to form a furnace wall, and the side wall material will be in direct contact with the electrolyte melt. In addition, the oxidation property of anodic gas is greatly enhanced with inert anode, and the strong permeation and destruction of potassium cryolite are added. Therefore, it is necessary to develop new antioxidation, anticorrosion, and resistant insulating sidewall lining materials and new cathode materials. The solution to the above problem may be to use oxide refractory containing Al_2O_3 [107] or metal material with oxidation resistance and corrosion resistance film [52] as the inner lining of electrolysis cell and use TiB_2 material without carbon as cathode [48].

3.7.3.4 Other Problems

With the decrease of electrolytic temperature, the influence of temperature on the density of aluminum liquid and electrolyte is not consistent, which causes the density difference between electrolyte and liquid aluminum to decrease and the electrolyte viscosity to increase. This brings difficulties to the effective separation of aluminum liquid and electrolyte and has a negative impact on current efficiency. With the decrease of electrolytic temperature, the conductivity of electrolytes decreases, which will be unfavorable to reducing the energy consumption of aluminum electrolysis. In addition, electrolyte volatilization and electrolyte interface properties may affect the electrolysis process.

Therefore, in order to guide the selection of low-temperature electrolyte system and the control of electrolytic process, it is necessary to study the physical and chemical properties of low-temperature electrolyte system and its control methods. This carried out the experimental research on low-temperature aluminum electrolysis of inert anode. It can not only reduce the corrosion rate of inert anode but also maintain the efficient and stable operation of electrolytic process.

3.8 Study on Inert Wettable Cathode

3.8.1 Advantages of Inert Wettable Cathode

The main advantage of the aforementioned wettable cathode or inert cathode is that the metal aluminum liquid and the cathode surface can be well wetted. It is possible to form a smooth and stable cathode interface (liquid aluminum interface) by maintaining only a layer of 3–5 mm thick liquid aluminum film, which can solve the problem of with carbon cathodes. The distance between anode and cathode can be shortened (from 4–5 cm of the existing industrial cell to 2–3 cm) by changing the structure of electrolysis cell (such as the diversion groove), so the potential of energy saving is huge [109, 110]. The inert cathode is the necessary foundation for the successful application of inert anode and the realization of energy-saving and environmental protection goals in aluminum electrolysis process (Table 3.3). In addition, the wet inert cathode can be directly used in the current electrolytic cell. Therefore, the alumina precipitate in the cell is not easy to deposit on the cathode surface, the cathodic current distribution is more uniform, and the pressure drop at the bottom of the furnace is reduced. Due to the good wetting of molten aluminum and the surface of the cathode, the wave peak caused by the fluctuation of molten aluminum can be weakened, and the height of molten aluminum cathode about 20 cm can be reduced appropriately, or all kinds of interference of production operation to the stability of magnetofluid in the electrolysis cell can be alleviated. The current efficiency can be improved at the same pole distance. The good wettability of aluminum liquid and inert cathode can reduce the permeation and destruction of electrolyte and sodium metal to the cathode and increase the life of electrolytic cell.

3.8.2 Requirements and Research Situation of Inert Wettable Cathode

Because the cathode of the aluminum reduction cell is submersible and its surface is always covered with aluminum liquid layer, it works slightly better than the anode. Ideal inert wettable cathode should be satisfied: (1) good wetting with molten aluminum; (2) refractory to high-temperature fluoride molten salts and molten aluminum and resistant to corrosion and penetration; (3) good electrical conductivity; (4) good mechanical properties, wear resistance, and thermal shock resistance at high temperature; (5) be able to combine well with the substrate to prevent electrolyte penetration; and (6) easy to process and form, convenient for large-scale production, wide source of raw materials, low cost of production, installation, and construction application.

The borides, carbides, silicides, and nitrides of the IV–VI para-group transition metals in the periodic table are commonly referred to as refractory hard metal (RHM). In the 1950s, the British Aluminum Company found that TiB_2 as one of RHM has good wettability with molten aluminum. It is envisaged that TiB_2 and other compounds can be used as inert and wettable cathode materials for aluminum reduction cells. It was later found that the borides and carbides of RHM, especially Ti and Zr had high melting point, high hardness, good electrical and thermal conductivity, and good wettability with molten metal. It can resist the corrosion and penetration of molten aluminum and cryolite-alumina molten salt and has the main properties required by inert and wettable cathode materials. However, these compounds are brittle and have poor thermal shock resistance. Table 3.14 lists the related physicochemical properties of TiB_2 and ZrB_2. The physical properties of TiB_2 and ZrB_2 are not different. The latter is more expensive than the former, as shown in Table 3.14. Therefore, TiB_2 ceramics and composites are mainly used as inert and wettable cathodes [111–113].

Inert wettable cathode materials based on TiB_2 have rapidly developed over the past few decades. A variety of TiB_2 inert and wettable cathode materials have been prepared by different methods, but there are three main types:

1. The pure TiB_2 is made into ceramic material with the shape of plate, rod, tube, etc.
2. Preparation of TiB_2-C composite cathode material from TiB_2 and carbon raw material.
3. Coating of TiB_2 on the surface of carbon cathode.

Table 3.14 Physical properties of TiB_2 and ZrB_2

Chemical compound	Melting point (°C)	Density (g/cm^3)	Resistivity/μΩ m		Thermal conductivity (W/m C)	Linear expansivity (°C^{-1})	Modulus of elasticity 25 °C (GPa)
			25 °C	1000 °C			
TiB_2	2850–2980	4.52	0.09–0.15	0.60	24–59	4.6×10^{-6}	253–550
ZrB_2	3000–3040	6.09–6.17	0.07–0.166	0.74	24	5.9×10^{-6}	343–491

3.8.3 TiB_2 Ceramic Wettable Cathode Material

The sintering properties of TiB_2 are poor, it is difficult to obtain TiB_2 materials with relative density more than 95% by pressure less sintering. Usually, TiB_2 ceramics with high density can be obtained by hot pressing sintering or cold pressing sintering with sintering aids. The relative density of hot-pressed sintered TiB_2 ceramic cathode materials can reach 95%–100%, but the cost is high, and it is difficult to prepare complex shape materials. The cost of cold pressing sintering is relatively low. Sintering aids, such as transition metal Co, Ni, and Cr, are added. Using TiB_2 ceramics as wettable cathode material exposes serious problems in the process of application. First, electrolytes and molten aluminum are easily permeated into the pores of ceramic materials during electrolysis, which corrodes the solid grain boundaries, resulting in a serious weakening of the adhesion force, cracks and failure, and damage after a relatively short time of use. Although intergranular corrosion can be weakened by using high purity TiB_2 as raw material, its cost is three or four times that of industrial pure TiB_2, and the cost is extremely high. Second, the linear expansion coefficient of TiB_2 ceramic material and cathode carbon block is quite different, and the thermal shock resistance is extremely poor, which makes it difficult to combine well with the matrix. TiB_2 is prone to brittle fracture, which is easy to rupture in the high-temperature environment of aluminum electrolysis. In addition, a large amount of TiB_2 powder is required to make ceramic materials. Because of the high price of TiB_2 and the above reasons, the TiB_2 ceramic cathode has not been applied in industry, only in laboratory research [114].

In 1957, Norton Company produced hot press-sintered TiB_2 bar for Reynolds Metals Company (RMC). The Reynolds connects the TiB_2 rod to the cathode steel bar and extends it through the lining through the bottom of the electrolyzer into the electrolytic cell and contacts with the molten aluminum to reduce the pressure drop at the bottom of the furnace. After 6 months of test in 68 kA electrolyzer, it was found that the hot press-sintered TiB_2 bar ruptured seriously and accompanied by intergranular corrosion. The raw materials of TiB_2 produced by carbothermal method contain a small amount of impurities, such as C, 0, and Fe. Most of these impurities are concentrated on grain boundaries. With the progress of aluminum electrolysis, electrolytes, sodium, and aluminum liquid permeate into the intergranular of the cathode material prepared from the TiB_2 powder, resulting in the rupture of the TiB_2 cathode material [114].

In the 1970s, Pittsburgh Plate Glass Corporation (PPG) developed a high-purity TiB_2 powder produced by non-carbothermal method. This kind of TiB_2 has perfect grain structure and pure grain boundary. It is considered to be a good raw material for wettable cathode. However, the preparation cost is too high. Moreover, the problems of high brittleness and poor thermal shock resistance of TiB_2 cathode materials have not been solved [115].

Many methods have been made on the application of TiB_2 ceramics as the cathode of aluminum electrolysis. In the past, TiB_2 was formed into plates, rods, tubes and grills, which were protruded to the aluminum solution. This were also used in the form of pieces and blocks on the surface of the cathode carbon block. Several fixed schemes shown in Fig. 3.28 are presented [116]. It is difficult to solve the problem of fixing TiB_2 ceramic cathode materials on carbon matrix in practical application. All kinds of methods have not been successful, and there has always been a problem of material breakage.

3.8.4 TiB_2-C Composite Wettable Cathode Material

The application of adding carbon material and TiB_2 to prepare composite cathode material reduces not only the amount of TiB_2 raw material but also the cost of TiB_2 wettable cathode material. However, it also improves the thermal shock resistance and mechanical strength of the material. It is easy to be prepared on a large scale and will not affect the electrical conductivity of the material.

TiB_2-graphite (TiB_2-G) composite cathode material for aluminum electrolysis was developed by Great Lakes Research Corporation (GLRC) in 1980 [117]. This kind of material has good thermal shock resistance, low solubility in liquid aluminum, and good corrosion resistance, completely wetting with aluminum liquid. The sample after placing aluminum liquid has been detected by SEM, from its surface to the inside of the material. The sample is about 1 mm infiltrating a layer of metal aluminum to form a so-called barrier layer that protects the material below [118]. From 1985 to 1986, GLRC cooperated with RMC to investigate TiB_2-G composite

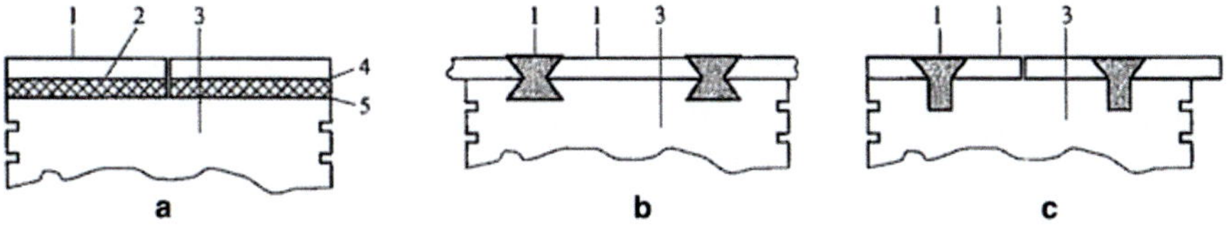

Fig. 3.28 The original structure of RHM fixed on cathode matrix. (**a**) Form of hot pressing or cement bonding; (**b**) key button form; (**c**) pin form. (*1*) Original RHM; (*2*) transition layer; (*3*) cathode matrix

wettable cathode materials. The TiB_2-G materials contained about (30%–40%) TiB_2 and were screened for different shapes of TiB_2-G materials. The "mushroom"-shaped TiB_2-G cathode component was selected [119]. In 1991, industrial tests were carried out on two 70 kA prebaked aluminum reduction cells in Kaiser Mead Smelter. Tabereaux [120] described the relevant technological conditions and operating process, and the experiment was carried out for 4–5 months. The results showed that the acd could be reduced by 2–2.5 cm, lower energy consumption than contrast cell (operated according to tradition process). However, the most difficult problem is the fracture and breakage of TiB_2-G cathode components, which results to failure to achieve the target of reducing energy consumption. After 12 days of startup, the TiB_2-G component is damaged and then gradually broken. Manufacturing defects are one of the reasons for the fracture of TiB_2-G cathode components. The mechanical pressure caused by aluminum outlet and pole exchange on TiB_2-G cathode components is also an important reason for the breakage of TiB_2-G cathode members and the effect of electrolytes in high-temperature molten salt environment. Both of them contributed to the damage of the TiB_2-G cathode structure. With years of research and industrial experiments, TiB_2-G cathodes have great application prospects. However, there are still many problems to be solved.

In fact, in addition to manufacturing defects and mechanical impact, the penetration and erosion of sodium and electrolytes are important reasons for the damage of TiB_2-C composite wettable cathode materials. In the electrolysis process, the cathode material is deformed and expanded under the action of sodium and electrolyte osmotic erosion, resulting in the damage and fracture of the material. Many researchers have studied the effect of sodium and electrolyte on TiB_2 composite wettable cathode material and its influencing factors [121, 122]. The results show that the wettability between liquid aluminum and cathode is not ideal when the content of TiB_2 inTiB_2-C composite cathode is 10%–30%. When the TiB_2 content is 40%–70%, the material has good wettability with the liquid aluminum. When the content of TiB_2 is greater than 70%, the material is completely wetted with the liquid aluminum [29, 123].

Cheng Geng et al. [124] developed a kind of integrated TiB_2-C composite cathode in China. A layer of TiB_2/carbon composite material was prepared by using the vibratory forming equipment of producing cathode carbon block in carbon plant to pressurize the original cathode carbon block. The experiment of TiB_2-C composite cathode in electrolytic cell shows that the TiB_2-C composite cathode can save energy in the current cell [125].

3.8.5 *TiB_2 Wettable Cathode Coating Material*

TiB_2 cathodic coating is a practical application of inert and wettable cathode materials for aluminum electrolysis. Because it combines the form of coating with the carbon cathode, it can be used either as an inert wettable cathode in a new type of electrolyzer or in an existing electrolyzer to improve the wettability of liquid aluminum with the surface of the cathode. By blocking or delaying the permeation of sodium and electrolyte, the working condition of the electrolytic cell is improved, the pressure drop at the bottom of the furnace is reduced, and the current efficiency is improved.

The concept of TiB_2 coating was first proposed by Martin Mtinia Company of the United States, that is, using the excellent wettability of TiB_2 material and aluminum liquid, using resin as binder, it is coated on the surface of carbon cathode of industrial aluminum electrolysis cell. Since then, the extensive research has been carried out on TiB_2 coatings, especially on corrosion resistance and wear resistance mechanism, properties, preparation technology, etc. In Europe, America, Australia, and China, TiB_2 coating technology has been reported. At present, there are two kinds of TiB_2 coatings. One is carbon-containing TiB_2 coating cathode material, such as TiB_2 (70%–90%) + carbon additive + resin binder, etc. The other is non-carbon TiB_2 coating cathode material, such as alumina sol TiB_2 coating.

The representative of the alumina sol TiB_2 coating is Moltech's Tinor coating, which has entered the industrial test. Experiments on electrolysis cell in collaboration with Hoogovens Company of the Netherlands show that it can reduce the penetration of Na to cathode carbon block. This can control the electrochemical corrosion of cathode carbon block [126]. The coating mainly uses the inorganic alumina sol as the bonding phase to bond the TiB_2 powder onto the carbon substrate to form a non-carbon TiB_2 coating cathode. Sekhgar [127] has made a commercial test on the cathode of non-carbon TiB_2 coating in 158 kA aluminum reduction cell. The results show that the coating is very hard and has excellent resistance to sodium permeability and wettability of molten aluminum. The results show that the current distribution of the coating tank is uniform and the cathodic pressure is reduced. In addition, there is no damage caused by sodium permeation. The coating can be used in both existing aluminum reduction cells and new aluminum reduction cells. The laboratory study and industrial test of colloidal alumina reinforced TiB_2 coating were carried out by Oye [128]. The coating has good wear resistance and aluminum wettability, which can block the permeability of sodium. However, the structure and properties of alumina sol and matrix carbon materials differs greatly, which will increase the resistance of cathode to a certain extent, and the durability of adhesive properties in the harsh environment of aluminum reduction cell needs to be further proved.

The typical representative of carbon adhesive coating is Comalco Company of Australia, which has been working on the research of carbon adhesive coating since 1987. It has developed a diversion electrolytic cell for the application of this coating. After more than 10 years of research and development from 1987 to 1998, the company has set up 25 diversion electrolytic cells with an electric current of 90 kA. The bottom of the cell is a TiB_2 coating cathode with both sides tilting inward, the carbon anode of the corresponding slanting bottom is used, and the central part of the trough is a polyaluminum trench. The liquid aluminum precipitated on the cathode can be pooled to periodically output aluminum. The thickness of the liquid aluminum layer on the cathode is 3–5 mm, and the polar distance is 2.5 cm (common trough is 4–4.5 cm). In order to maintain thermal balance, the anode current density increased to 1.15 A/cm^2 and the current increased to 120 kA, thus increasing the output by 40% and reducing the energy consumption to 13,200 kWh/t-Al. Later, the company improved the formulation of the coating material, and the industrialization of the TiB_2 coating were successful [129].

The research on TiB_2 wettable cathode coating materials in China started at the same time in the world, and industrial electrolysis experiments were carried out on the basis of laboratory research. Li Qingyu has developed a new type of TiB_2 cathode coating material cured at room temperature. This can save the high-temperature curing process of the coating and can be calcined and carbonized at room temperature for a certain time (about 24 h), thereby saving the heating equipment needed for high-temperature curing. The coating construction time is shortened and the application cost is reduced, which is beneficial to the popularization and application of TiB_2 cathodic coating technology. The results of industrial application (color iv) show that the solidification and carbonization effect of TiB_2 electrode coating material cured at room temperature is good. The electrolysis cell starts smoothly, the bottom of the furnace is clean, the furnace chamber is regular, the pressure drop of the bottom of the furnace is reduced, and the penetration of sodium and electrolyte is obviously slowed down. It has remarkable effect on prolonging the life of electrolytic cell [130]. In addition to the two main TiB_2 coating materials mentioned above, there are many other ways to prepare TiB_2 coating materials, including electrodeposition of TiB_2 coating, plasma spraying of TiB_2 coating, laser spraying of TiB_2coating, vapor deposition of TiB_2 thin layer, and self-propagating TiB_2 thin layer. Because of complex equipment, large area construction difficulties, and high cost, these methods have not been reported in industrial aluminum reduction cells. It is even more difficult to meet the performance and cost requirements of wettable cathode materials used in new aluminum reduction cells.

3.9 The New Type of Electrolysis Cell Based on Inert Electrode (Anode and Cathode)

Based on the inert anode and wettable cathode mentioned above, many kinds of new aluminum electrolytic cells have been proposed or designed for energy saving and environmental protection in aluminum electrolysis process. This section introduces some typical electrolytic cells.

3.9.1 Electrolytic Cell with Inert Anode

Only inert anode is used in the electrolytic cell. The carbon anode of Hall-Heroult aluminum reduction cell is replaced by inert anode, and the other parts are basically unchanged. The cermet anode 6 kA electrolyzer tested by Alcoa et al. [60] is a typical example. The advantage of this kind of electrolytic cell is that it is convenient to transform the existing Hall-Heroult aluminum reduction cell and the investment is relatively small. Its disadvantage is that the effective electrolytic area is small and the space utilization ratio of the electrolytic cell is low. Thus, it is difficult to increase the aluminum production per unit volume by increasing the effective electrolytic area of the electrode. In addition, it is difficult to reduce the energy consumption effectively by reducing the pole distance, because of the failure to solve the problems such as the reduction of current efficiency at low pole spacing caused by the instability of aluminum solution in the electrolytic cell. In addition, when the inert anode is used in electrolysis, the electrolytic cell needs to run at a higher alumina concentration. The precipitation of alumina is serious, which affects the normal process of electrolysis. The reversible decomposition voltage is higher than that of the carbon anode. Therefore, the energy consumption of the electrolytic cell with the same polar distance will be higher than that of the current Hall-Heroult aluminum electrolyzer [10].

3.9.2 *Electrolytic Cell Using Wettable Cathode Alone*

Carbon anode is still used in anode, and only aluminum cell with wettable cathode is used. In addition to the Hall-Heroult aluminum electrolysis cell coated with wettable TiB_2 material on the surface of the cathode carbon block, there are a variety of new aluminum electrolysis methods to improve the cathode.

In cooperation with Hefei Aluminum Plant, Central South University of China has carried out the industrial test of 28, 60 kA electrolytic cells coated with TiB_2 cathode. The results are satisfactory. The current efficiency is increased by 2.4, and the energy consumption is reduced by 340 kwh/t-Al. The life of the electrolytic cell is greatly prolonged, and the operation is not damaged for more than 5 years. The cost of TiB_2 coating is much lower in China than that in other countries [110].

3.9.3 *Mushroom Cathode Electrolysis Cell*

There are wettable cathodes with mushroom shape, coated with wettable material and parallel to anode bottom, and the root is connected with cathode guide rod through cathode carbon block, as shown in Fig. 3.29. The liquid aluminum precipitates on the surface of the wettable cathode and flows into the bottom of the trough. There is only a thin layer of liquid aluminum on the surface of the cathode, which can reduce the polar distance properly, and the cathode can play a certain role in maintaining the stability of the liquid aluminum. The problem with this type of electrolyzer is that the cathodic material has insufficient corrosion and impact resistance. This is prone to melting or fracture [112, 120].

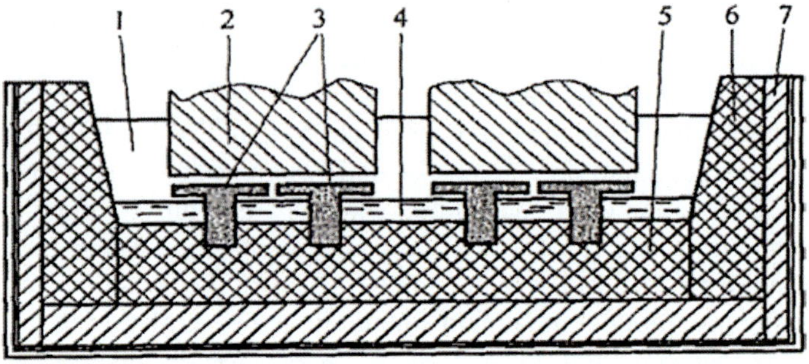

Fig. 3.29 The electrolytic cell with "mushroom" wettable cathode installed at bottom of tank [112]. (*1*) Electrolyte solution, (*2*) anode, (*3*) wettable cathode member, (*4*) aluminum metal, (*5*) cathode carbon block, (*6*) lateral carbon block, (*7*) thermal barrier layer

3.9.4 Conducting Aluminum Electrolytic Cell with Carbon Anode (Flame Diversion Trough)

Another type of electrolytic cell with a wettable cathode is a trench, which has been widely preferred for many years. Since the 1970s, there have been many patents on grooves. The characteristic of the trench is that the surface of the carbon cathode is coated with wettable coating of TiB_2, the surface of the cathode is 2° or more inclined due to the good wetting of the liquid aluminum to the surface of the cathode. This makes liquid aluminum to flow along the slope into the bottom grooves (polyaluminum trenches), on the premise of obtaining higher current efficiency, the pole distance can be controlled between 1.2 and 2.5 cm. According to the structure and distribution of polyaluminum trench, the channel can be classified into two types: monopolymer aluminum groove and polyaluminum grooves. The most typical monopolymer channel structure, which is relatively simple and can be implemented in the process of reconstruction of damaged electrolyzer or new electrolytic cell, as shown in Fig. 3.38. The difficulty is relatively low. Therefore, it has considerable attraction. From 1987 to 1998, Comalco Company [129] of Australia has been researching and developing flame diversion trough. Using the structure of the trench shown in Fig. 3.30, 25 test tanks with current of 90 kA have been established.

At the bottom of the groove is a TiB_2-coated cathode, the sides of which are inclined inward. The carbon anode with a suitable bottom surface is used. The polyaluminum groove is located in the middle of the groove bottom. The thickness of the liquid aluminum layer on the cathode is 3–5 mm and the polar distance is 2.5 mm. In order to maintain the thermal balance, the current has been increased from 90 kA to 120 kA. The anode current density has been increased to 1.15 A/cm^2, and the aluminum production has been increased by 40%, and the energy consumption is 13,200 kwh/t-A1.

Georges [132] described the structure of monopolymer aluminum trench in the patent, as shown in Fig. 3.31. The cathode of the electrolytic cell tilted inward from both sides, forming a polyaluminum trench at the bottom of the trough, and coated

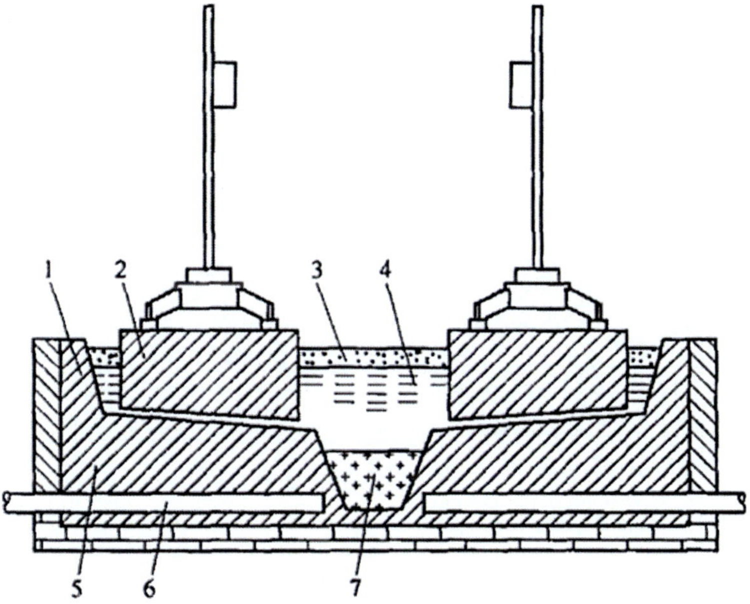

Fig. 3.30 The typical structure of channel of monopolymer aluminum trench [129]. (*1*) lateral carbon block, (*2*) carbon anode, (*3*) crust, (*4*) electrolytes, (*5*) cathode carbon block, (*6*) cathode steel bar, (*7*) aluminum liquid in polyaluminum ditch

wettable material on cathode surface. A fixed shell with a cathode in the cell (referred to as inner groove shell). The insulation material (such as refractory brick) is used to separate it from the external slot body to insulate the inner groove shell from the rest of the cell body. In addition, it provides a space to control the temperature of the inner groove shell by injecting heating or cooling gas into it. Especially at start-up, this method can be used to preheat the cell. The inner groove shell is used to ensure the uniform distribution of the current in the cathode carbon block and can be separated from the electrolytic cell as a whole. It is easy to replace. The cathode guide rod can be inserted into the cell body from both sides and connected to the inner groove shell. It can be introduced from the vertical opening at the bottom of the groove, where the cathode guide rod is located at the geometric center of the cathode block at the bottom of the slot and welded to the inner groove shell. Although the structure of this kind of diversion tank is complex, the design is quite advanced, especially for the design of thermal balance of aluminum reduction cell. In future applications, if it is a new aluminum factory, they will have considerable attraction.

V. de Nora [133] described the polyaluminum channel structure shown in Fig. 3.32 in the patent. This type of cell is characterized by the transverse alignment of cathode blocks across the bottom of the cell into many grooves (polyaluminum ditch), the liquid aluminum flows down the cathode slope to the recess on either

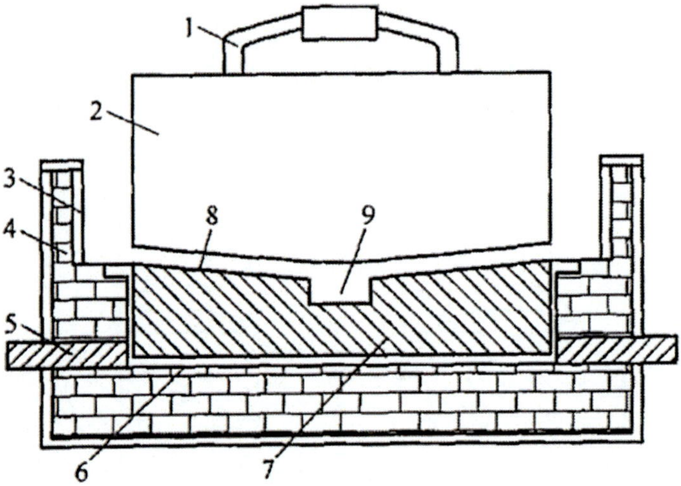

Fig. 3.31 Schematic diagram of channel structure of monopolymer aluminum trench [132]. (*1*) Anode steel claw, (*2*) anode, (*3*) silicon carbide layer, (*4*) refractory brick, (*5*) cathode guide rod, (*6*) inner groove shell, (*7*) cathode block, (*8*) cathode coating, (*9*) aluminum sulcus

side. The surface of carbon cathode is coated with TiB_2 wettable coating, which makes it wettable to molten aluminum, and the cathode guide rod is still steel material. The patent also describes a schematic diagram of the cathode blocks, in which the polyaluminum trenches are V, U, trapezoid, and rectangle, bonding between cathode carbon blocks with tamping paste.

The structure of this channel is more complex, and many grooves are formed at the bottom of the groove. If there is no other channel to collect the liquid aluminum, the extraction from aluminum will be more troublesome. In addition, it is necessary to produce cathode carbon blocks and anode carbon blocks with special structure. It is difficult to adjust the pole distance during electrolysis and to maintain the uniform pole distance when the anode is replaced.

The properties of current guide cell with carbon anode is better than that of common aluminum reduction cell. However, it is expected to reduce the pole distance and energy consumption by a large margin. To solve the disadvantages of the current aluminum reduction cell and achieve the goal of greater energy saving, increasing production, and improving environmental protection, it is necessary to develop a new type of aluminum reduction cell with inert anode and wettable cathode.

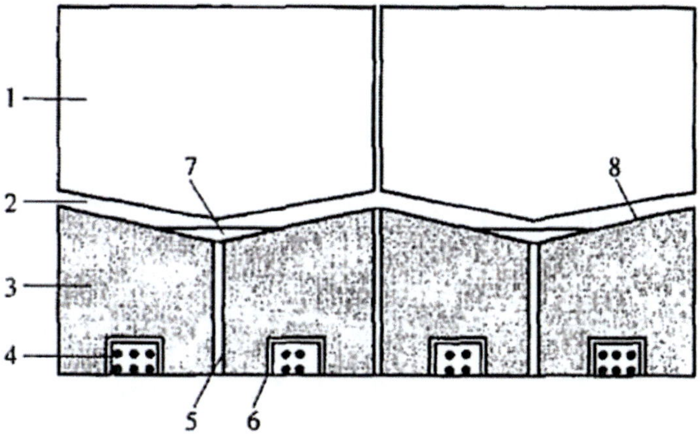

Fig. 3.32 The polyaluminum trench structure [133]. (*1*) Anode, (*2*) interpolar electrolyte, (*3*) cathode block, (*4*) cathode steel rod, (*5*) tamping, (*6*) cast iron, (*7*) aluminum liquid in polyaluminum ditch, (*8*) titanium boride coating

3.9.5 Electrolytic Cells that Combine Inert Anode and Wettable Cathode

3.9.5.1 Inert Anode Diversion Trough for Monopolymer Aluminum Ditch

Georges [132] presented a channel with inert anodes (Fig. 3.33). This channel differs from that in Fig. 3.31 in the use of inert anodes, for example, Ni-Fe-AI alloy or Ni-Fe-Al-Cu alloy with oxide protective layer on the surface, other structures are basically similar.

3.9.5.2 An Inert Anode Electrolysis Cell for Polyaluminum Channel

V. De Nora [133] presented a polyaluminum channel with inert anodes, as shown in Fig. 3.34. Inert anodes may be made of alloys or ceramics coated with oxides or fluorides as protective coatings, etc. The anode gas is discharged along the opening in the middle of the anode, and the rest is similar to that in Fig. 3.32.

3.9.6 An Inert Anode Aluminum Electrolysis Cell with Complex Structure

As shown in Fig. 3.35, V. de Nora [134] presented a new type of aluminum electrolysis cell with complex structure in its patent. The electrolysis cell uses both inert anode and wettable cathode, which can be classified as a diversion cell. The coating

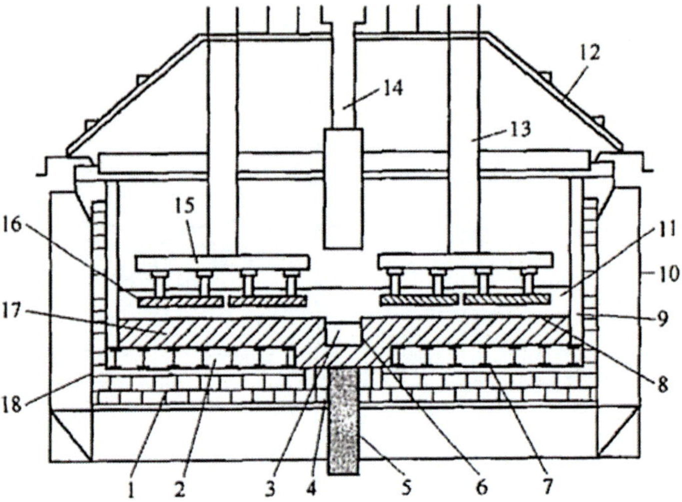

Fig. 3.33 The structure of inert anode diversion cell in monopolymer aluminum ditch [132]. (*1*) Refractory bricks, (*2*) air hole, (*3*) aluminum liquid, (*4*) vertical opening, (*5*) cathode guide rod, (*6*) aluminum sulcus, (*7*) trestle, (*8*) cathode slope, (*9*) side lining, (*10*) external groove shell, (*11*) electrolytes, (*12*) groove cover, (*13*) anode guide rod, (*14*) crust breaking and baiting device, (*15*) shunt device, (*16*) anode, (*17*) cathode block, (*18*) inner groove shell

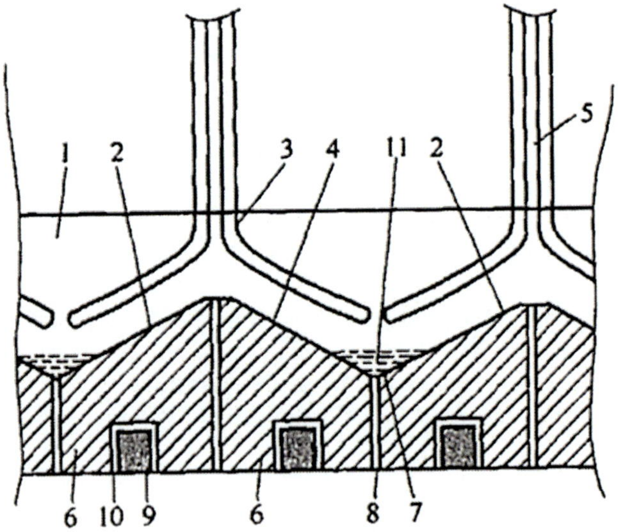

Fig. 3.34 The structure of inert anode diversion cell in polyaluminum channel [133]. (*1*) Electrolytes, (*2*) cathode slope, (*3*) anode, (*4*) titanium boride coating, (*5*) anodic gas channel, (*6*) cathode carbon block, (*7*) aluminum sulcus, (*8*) tamping, (*9*) cathode steel rod, (*10*) cast iron, (*11*) aluminum liquid

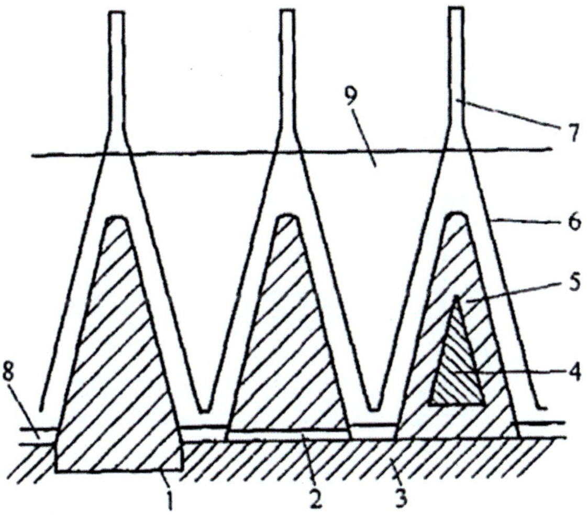

Fig. 3.35 Inert anode guide cell with complex structure [134]. (*1*) A cathode fixed to the bottom of cell, (*2*) binder, (*3*) aluminum liquid corrosion-resistant layer, (*4*) cast iron, (*5*) cathode block, (*6*) anode, (*7*) anode opening, (*8*) aluminum liquid, (*9*) electrolyte

on cathode surface has good conductivity and wettability of aluminum liquid, and it has a good protective effect on cathode carbon block. The surface is used for coating the wedge-shaped cathode of the boron-coated titanium; the cathode can be fixed on the bottom of the groove by the way of the bottom of the implanting groove. The cathode carbon block can be combined with the bottom of the groove through the bonding, or cast iron is added inside the cathode carbon block so as to be sunk to the bottom of the groove. In addition, inert anode, such as metal, alloy, or ceramic, is inclined in herringbone, parallel to the cathode surface, and has an opening for anode gas emission. The pole distance is controlled within 15–20 mm. The electrolytic precipitated aluminum flows down the cathode slope to the bottom of the trough. This kind of electrolytic cell is actually an improvement of mushroom cathode and faces the problem of cathode service life. The wedge cathode will face the problem of fracture and coating peeling under electrolytic conditions.

1. Bipolar electrode vertical electrolytic cell: As shown in Fig. 3.36, this battery structure is similar to a magnesium battery. Each electrode serves as an anode on one side and a cathode on the other side, and each pair of electrodes is combined into an electrolytic cell, which is then connected in series. The current flows from one end of the battery to the first anode, then through the electrolyte to the cathode surface of the next electrode, and finally to the last cathode at the battery end. Therefore, electrolysis can be performed at a lower current and a higher battery voltage, which makes it easier to transfer current. Its pole distance can be controlled in a small range. Thus, the whole groove can be designed more compact and has a higher output rate [135].
2. The vertical electrolytic cell has unipolar electrode, except that the electrodes at both ends are conducting on one side and the electrodes in the middle are conducting on both sides. However, the polarity is the same. Each combination of two electrodes can be regarded as an electrolytic cell. The difference between the

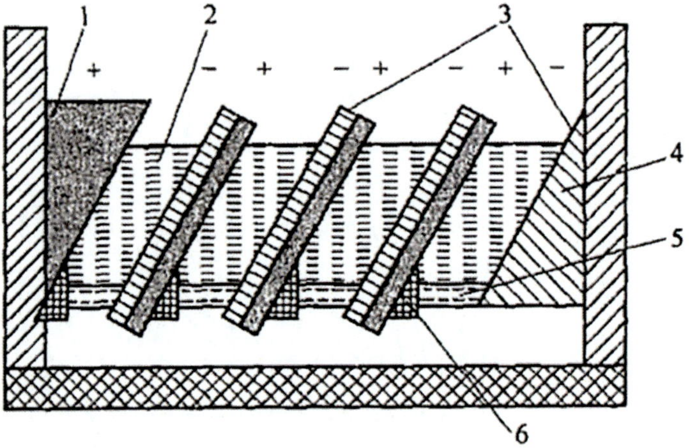

Fig. 3.36 The vertical electrolytic cell with bipolar electrode [135]. (*1*) Anode, (*2*) electrolytes, (*3*) bipolar electrode, (*4*) cathode, (*5*) aluminum liquid, (*6*) inert insulating material

two electrodes is that the electrode combination is parallel rather than in series. Various unipolar electrode electrolytic cell is described in V. De Nora's series of patents. Figure 3.37 is one of the cells using a vertical inert cathode and a vertical inert anode [136]. The anode is suspended from the anode busbar through the guide rod, and the bottom of the cathode is fixed at the bottom of the groove and connected to the cathode bus at the bottom of the groove. The effective electrolytic area of this kind of cell is much larger than that of the electrolytic cell. This only uses the bottom of the cell as the cathode. Under the condition of ensuring the same output rate per unit area, the capacity of the electrolytic cell is significantly improved. Due to the small distance between the poles, there are only upward fluids between the poles. On the cathode and anode side, there is room for the downward flow space, which can be used as the place of discharge.

Bipolar electrode electrolysis cell and unipolar electrode electrolyzer have their own disadvantages. One of the most prominent problems in bipolar electrolytic cells is that the bypass current is more serious. Part of the current flows directly from the first anode through the electrolyte, the aluminum solution, or the sidewalls to the last cathode, instead of flowing through the middle electrode, reducing current efficiency. If the electrolyte and liquid aluminum are separated from each pair of electrodes, there will be no bypass current (only a small amount of current flows through the battery wall). However, this will cause great difficulties for the bait and output aluminum. The current distribution of the unipolar electrode electrolyzer will be more uniform, and the voltage drop in the electrode and busbar will be greater, which reduces the power efficiency. Beck [137] presented the calculation method of the current efficiency of unipolar electrode and bipolar electrode. They believed that the current efficiency of the two kinds of electrolytic cell was similar. In addition, when these two electrolytes are used for electrolysis, it may be difficult to transfer mass when electrolysis temperature is

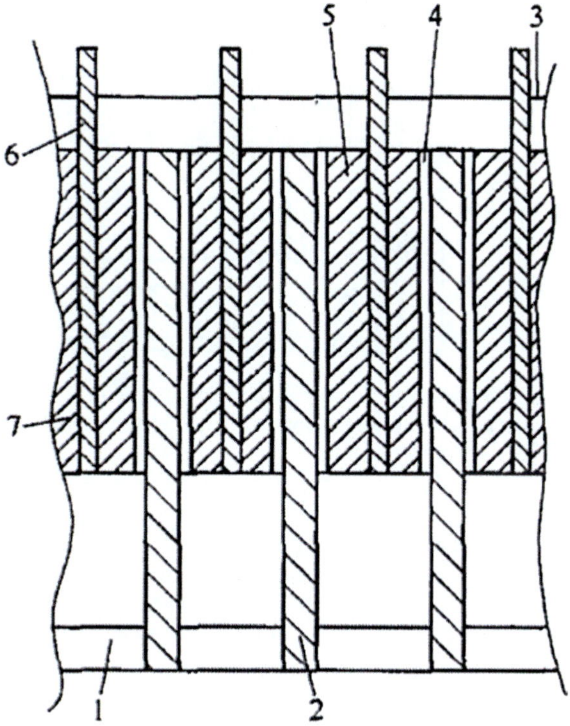

Fig. 3.37 The vertical electrolytic cell with unipolar electrode [136]. (*1*) Aluminum liquid, (*2*) cathode, (*3*) electrolytes, (*4*) polar spacing, (*5*) anode, (*6*) guide rod, (*7*) inert material

low and viscosity is high. This may cause aluminum liquid to suspend in electrolyte and make "secondary reaction" serious.

3.9.7 Slurry Electrolyzer

The low-temperature electrolysis is beneficial to reducing the corrosion and thermal shock of electrolyte melt to inert anode, and it is the only way to develop inert anode. However, when electrolyzed at low temperature, the alumina has low solubility and decreased dissolution rate. When the Al_2O_3 supplement is insufficient, the anode potential will increase as the concentration of alumina near the anode decreases. In addition, the reaction between the oxide on the anode surface and the electrolyte will intensify, and even catastrophic corrosion will occur [82] (dissolution of anode metal and electrochemical decomposition of oxide).

In order for the electrolysis to proceed smoothly, there must be undissolved alumina in the electrolyte to replenish the alumina consumed near the electrode in time. This can keep the current density within a reasonable range. However, it can easily lead to a large amount of alumina precipitation [101]. In order to solve the above problem, Beck proposed the slurry electrolysis cell shown in Fig. 3.38 in the

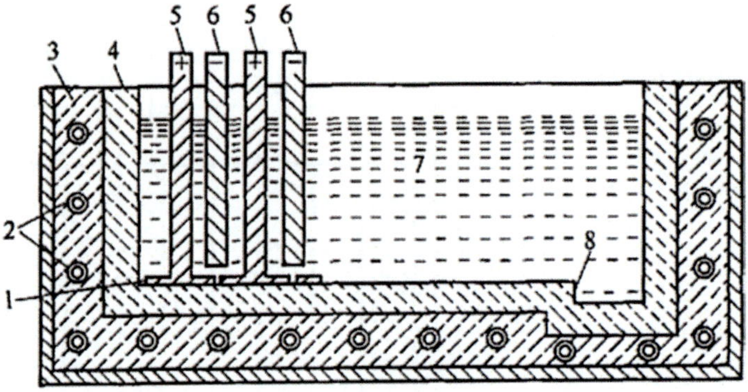

Fig. 3.38 Structure diagram of slurry electrolytic cell [131]. (*1*) Anode bottom, (*2*) tank body cooling pipe, (*3*) keep warm material, (*4*) refractory lining, (*5*) anode, (*6*) cathode, (*7*) electrolyte melt, (*8*) aluminum sulcus

patent [131]. This electrolysis cell uses the vertical unipolar electrode, and the bottom of the cell is used as the anode. During the electrolysis process, the rising anode bubble can ensure that the Al_2O_3 is suspended in the electrolyte. The precipitated aluminum liquid is wrapped in the electrolyte and deposited in the polyaluminum ditch at the bottom edge of the trough as the electrolyte flows. The electrolytic temperature can be controlled by cooling tubes on the side wall and bottom of the cell. However, it is difficult to ensure that the aluminum solution will not be re-oxidized during the deposition process.

3.9.8 The Future Development of New Aluminum Reduction Cell

The great progress has been made in inert anode, wettable cathode, and new structure electrolyzer based on inert electrode. At present, new electrolytic cells using or combining inert anodes and wettable cathodes have been developed one after another. Although these electrolytic cells have the potential advantages of energy saving, environmental protection, and high production capacity. However, they still have disadvantages. Various electrolytic cell structures should be further optimized, and a series of engineering technical problems in electrode materials and electrolytic processes should be solved. In view of the future development of new aluminum reduction cells based on inert anode, wettable cathode, and the new electrolytic process, the United States and other countries already have a series of guiding documents, for example, "Technical guide for inert anodes" (1998) and "Current status of inert anode technology" of the United States (1999) [138].

3.10 Recent Research and Development of Aluminum Electrolytic Inert Anodes

At present, the research of inert anode has entered the engineering test stage in two material systems of alloy anode and cermet anode. The oxide ceramic inert anode has excellent electrolyte dissolution corrosion resistance and new oxygen infiltration corrosion resistance. However, it is difficult to connect with metal guide rod and so on because of its low conductivity and high brittleness. This kind of material has been gradually abandoned. In recent years, the research of inert anode has focused on the dynamic balance control of the formation and dissolution of the surface protective film of alloy anodes, the optimization of composition and structure of metal ceramic, the electrolytic corrosion mechanism and inhibition technology of inert anode, etc. Since 2007, this paper continues to review the latest research progress in alloys and cermet inert anodes for aluminum electrolysis.

3.10.1 Cermet

Samoylov et al. prepared cermet, [139] including mixing iron oxide powder Fe_2O_3 and nickel oxide powder NiO, pressing as ingot, thermal treatment of pressed forms in air at 1000–1200 °C, and subsequent grinding of sintered specimens. The production of cermet usually includes mixing $NiFe_2O_4$, NiO, and metal powder (Cu or Ag), then pressing to reach 60% of their theoretical density, and sintering it in an argon flow, heating up to 1350 °C for 4 h and holding at this temperature for 4 h. Finally, it must be cooled down to room temperature. They achieve a density of the specimens from 5.46 to 5.52 g/cm^3.

A two-step sintering process was adopted by Zhang et al. [140] to prepare $NiFe_2O_4$ inert anodes. For the production of $NiFe_2O_4$ spinel, Fe_2O_3 and NiO powders were heated at 1000 °C for 6 h. After crushing and sifting the oxide, they added $NiFe_2O_4$ nanopowder. The prepared mixture was then uniaxially compressed, and they sintered the nickel ferrite a second time at 1300 °C for 6 h. The addition of 0 to 40 wt% $NiFe_2O_4$ nanopowder significantly influenced the density, porosity, bending strength, and impact toughness. Inert anodes with 30 wt% nanopowder had the best compressive properties. Using this composition, they obtained the following results: density of 4.86 g/cm^3, porosity of 3.5%, bending strength of 42.47 MPa, and impact toughness of 3.31 J/cm^2.

Wu et al. [141] prepared nano-cermet for aluminum electrolysis, including precipitation, mold pressing and sintering process. In addition, the nano-cermet was characterized by transmission electron microscopy (TEM) and X-ray diffraction (XRD). The experiments and tests on corrosion and electrical conductivity show that the nano-cermet can be used as an inert anode of aluminum electrolysis. It can improve the mechanical properties and processing properties of inert anode cermet for aluminum electrolysis.

3.10.1.1 Regulation and Control of Sintering Atmosphere

For the most concerned $NiFe_2O_4$-based cermet in cermet inertia anode material, the ceramic matrix phase $NiFe_2O_4$ is a non-stoichiometric system. The content of Fe, Ni, and O plasma and the valence state of Fe elements vary with the change of oxygen partial pressure and temperature (Fig. 3.39) [142]. The sintering behavior and the material property after sintering strongly depend on the oxygen partial pressure of the system. When sintering is carried out at an appropriate partial pressure of oxygen, typically an Ar atmosphere or N_2 atmosphere with a fixed oxygen content, the $NiFe_2O_4$ retains both the cusp-type structure and the high oxygen vacancy concentration. The oxygen vacancies are important mass transfer sources in the sintering process. The migration of the air holes and the moving speed of the grain boundaries are increased, and the grain growth and the sintering densification are promoted. In addition, in the process of forming oxygen vacancy, part of Fe^{3+} is reduced to Fe^{2+}, and the Fe^{2+} first occupies the b site of octahedral. The increase of octahedral Fe^{2+} concentration increases the probability of electron jumping between Fe^{2+} and Fe^{3++}, which makes $NiFe_2O_4$ ceramics exhibit good electrical conductivity. There is no significant difference in the lattice parameters between $NiFe_2O_4$ ceramics sintered in air and N_2 atmosphere. However, there are significance in the grain size and density of sintered ceramics between them, as shown in Fig. 3.40 [143].

The effect of the sintering atmosphere was examined by Zou et al. [144]. 5Cu/(10NiO-$NiFe_2O_4$) cermet anodes were prepared by the cold-pressing sintering method in different atmospheres. Furthermore, the phase composition, microstructure, and mechanical properties were investigated. The results showed that 5Cu/

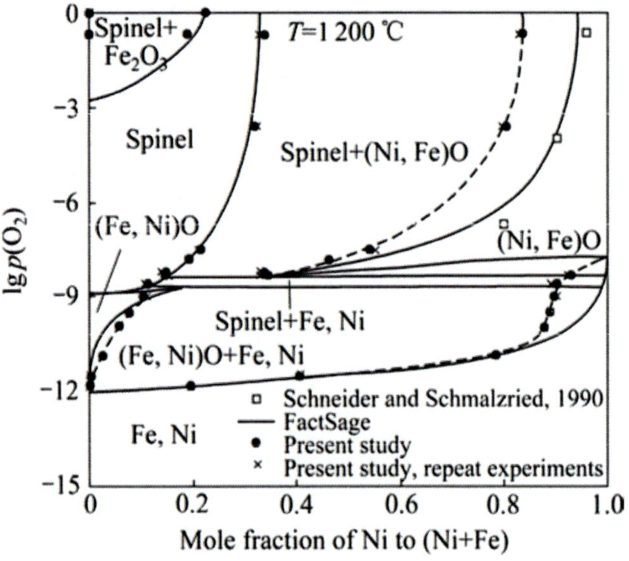

Fig. 3.39 Isothermal phase diagram of Fe-Ni-O at 1200 °C [142]

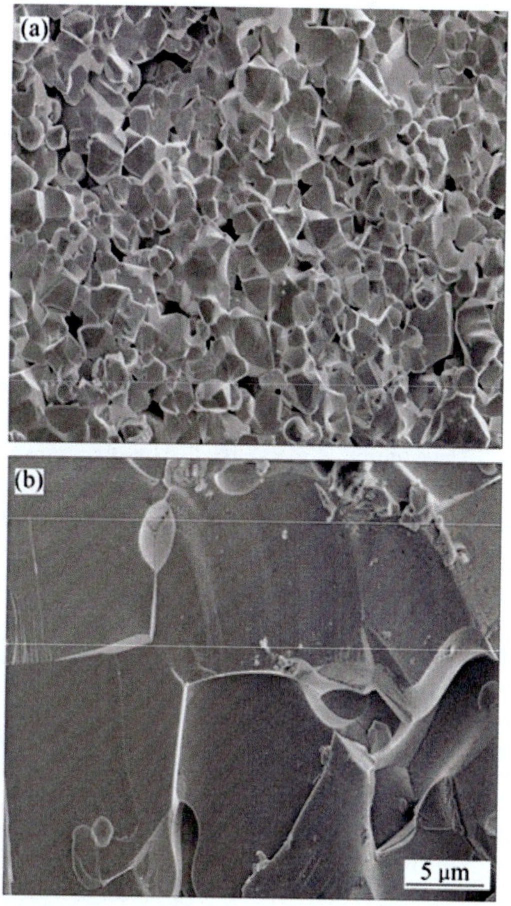

Fig. 3.40 Cross-sectional SEM images of $NiFe_2O_4$ ceramic sintered in (**a**) air and (**b**) N_2 [143]

($10NiO-NiFe_2O_4$) cermet materials can be obtained by sintering in a vacuum (oxygen content 0.02 ppm) or in atmospheres of Ar^+ with oxygen content 10 ppm, 200 ppm, 2000 ppm, and 10,000 ppm, respectively. The $NiFe_2O_4$ content in the ceramic material increased with the oxygen content in the sintering atmosphere and the Cu in the $NiFe_2O_4$ phase decreased. When the oxygen content in the sintering atmosphere was only 10 ppm, the grain size of $5Cu/(10NiO-NiFe_2O_4)$ cermet was 5.43 μm, and the bending strength reached 80.05 MPa at room temperature.

Then, Wang and Sun [145] examined the electrochemical behavior of $5Cu/(10NiO-NiFe_2O_4)$ cermet in low-temperature melting electrolytes saturated with aluminum melt electrolysis. Two anodic processes were observed at the cermet anodes: the dissolution of copper and evolution of oxygen. The surface activity of anodic films indicated the lack of surface oxide stability. The exchange current density of cermet anodes was 0.19 A/cm^2. The porous surface caused by the depletion of copper during anodic polarization could explain the oxygen evolution blocking on the surface of the cermet anode.

Liu et al. [146] examined the phase evolution of 17(Cu-Ni))-(10NiO-$NiFe_2O_4$) inert cermet anodes. A dense $NiFe_2O_4$ layer was observed on the anode surface and thickened with longer electrolysis time. In the newly formed dense ceramic layer, the NiO phase disappeared as a result of being swallowed by the $NiFe_2O_4$ phase. The metal phase was oxidized during the electrolysis process, and the dissolution rate of Cu is higher than that of both Fe and Ni.

3.10.1.2 Mix Doping

Adding metal element Ni to Cu can improve the wettability of Cu and the ceramic phase and increase the melting point of metal phase. The density of the metal phase was higher than that of metal phase Cu. At present, it has been widely used by researchers [147, 148].

The addition of metal, Ni also has the impact on its effect. For example, the relative density and conductivity of the samples with metal phase coated powder were higher than that of the simple mixed powder (Fig. 3.41) [149]. When sintered at 1300 °C, the relative density of the coated powder sample was 98%, and the

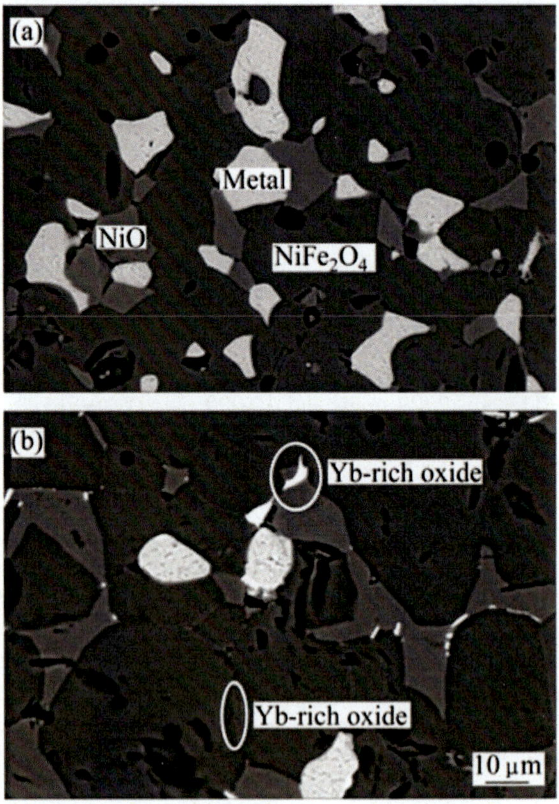

Fig. 3.41 SEM images of $NiFe_2O_4$-10NiO-based cermet with 17% (mass fraction) metallic phase [149]. (**a**) 95Cu + 5Ni mixture; (**b**) 95Cu-5Ni coated powder

conductivity was 78 s/cm at 960 °C. Compared that of the mixed metal powder sample, the porosity and the conductivity of the coated powder sample decreased by 1.3% and nearly increased by 20%, respectively.

It is found that BaO can react with these two phases to form new phases, such as $BaFe_2O_4$, $Ba_2Fe_2O_5$, $NiBa_3O_4$, etc.

1% BaO can significantly increase the relative density of $Cu/(NiFe_2O_4\text{-}10NiO)$ cermet, which is 96%. Compared with the relative density of undoped sample, it increases by 6%. Similar to the effect of BaO addition, the addition of Yb_2O_3 generated $YbFeO_3$ phase. When the doping content of Yb_2O_3 is 0.5%, the relative density of sintered samples at 1275 °C reached 95%, which increased by 2.0% compared with that of undoped samples. The bending strength and electrolytic corrosion rate of the materials were significantly reduced [150]. However, it is necessary to consider the pinning effect of new phase particles on grain boundary migration of $NiFe_2O_4$ phase. For example, the microstructure of $YbFeO_3$ phase distributed at the grain boundary was closely related to the doping content. When the doping content is greater than 2.0%, the $YbFeO_3$ phase particles gather gradually at the grain boundary to form a film (Fig. 3.42). The conductivity, mechanical properties, and corrosion resistance of the material decrease.

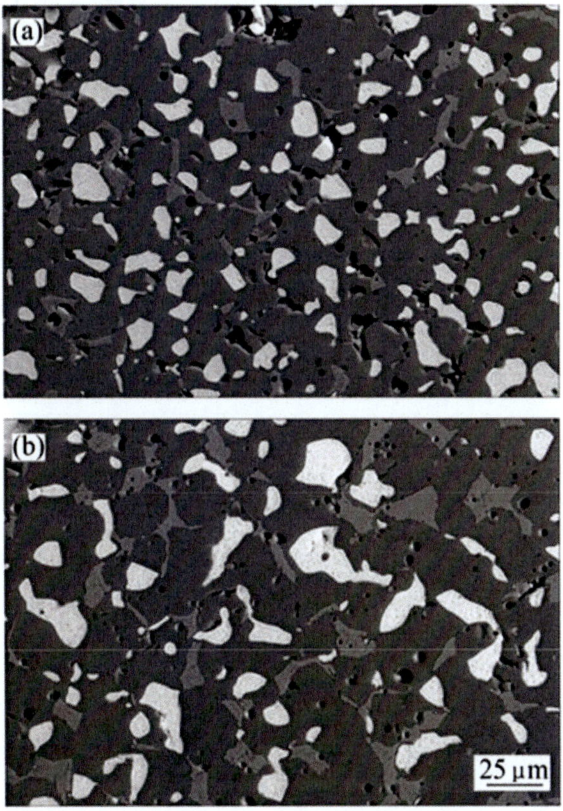

Fig. 3.42 SEM images of $10Cu/(NiFe_2O_4\text{-}10NiO)$ cermet sintered at 1300 °C [150]. (**a**) Undoped; (**b**) doped with 2.0% Yb_2O_3

He et al. [151] researched whether the addition of oxides can improve corrosion resistance. They added 1% BaO to $xCu(10NiO\text{-}NiFe_2O_4)$ cermet ($x = 5, 10, 17$) inert anodes. They observed severe corrosion of Cu during the experiments. Many pores appeared on the surface of the anode, and electrolyte infiltrated into the anode during electrolysis. Adding BaO to the cermet material is not conducive to corrosion resistance. It is because that BaO at the grain boundary of the anode accelerates the corrosion of the cermet.

Liu et al. [152] observed that the addition of TiO_2 and V_2O_5 to $NiFe_2O_4$ spinel could improve the electric conductivity and corrosion resistance because of the change of the microstructure of the $NiFe_2O_4$ grain boundary.

Du et al. [154] investigated the effect of MnO_2 additions on the early-stage sintering behavior and properties of $NiFe_2O_4$ ceramics. Adding 1.0 wt% MnO_2 to $NiFe_2O_4$ can improve the density. The maximum relative density is 93.63%, and the bending strength is 38.75 MPa.

Zhang et al. [153] studied the effect of addition of ZrO_2 on properties of nickel ferrite cermet. The results showed that the relative density slightly increased by 0.5 wt%. The decrease in corrosion resistance was no significant. However, the bending strength increased from 55.5 MPa (without ZrO_2) to 105.26 MPa. The higher the ZrO_2 content, the higher the corrosion rate.

Du et al. [154] reported that $NiFe_2O_4$ ceramic-based inert anodes were fabricated by a two-step cold-pressing sintering process. The bubble behavior of $NiFe_2O_4$ ceramic-based inert anodes was investigated in a two-compartment see-through quartz cell. The anodic overvoltage and potential decay curves on the inert anodes were measured by using the steady-state and current interruption techniques. The results showed that the electrolytic gas evolution for $NiFe_2O_4$ inert anodes, including bubble nucleation, growth, coalescence, growth again, migration, and escaping, lasted for 79 s, and the escaping bubble size was about of $\Phi 4$ mm × 2 mm. In the carbon anode, the gas was continuously released for 102 s, and the released bubbles were larger. When current densities were 0.6, 0.8, 1.0, and 1.2 A/cm^2, the anodic overvoltage of $NiFe_2O_4$ anodes were 0.189 V, 0.270 V, 0.309 V, and 0.359 V, respectively. After adding small amount of MnO_2, V_2O_5, and TiO_2, a minor reduction in anodic overvoltage of $NiFe_2O_4$ anodes could be obtained.

3.10.2 Alloy Inert Anode

The representative alloy anodes were Cu-Al [155, 156] and Cu-Ni-Fe alloys [157, 158] studied by Argonne National Laboratory, especially Moltech's Ni-Fe alloy [17]. The company had carried out 4 kA aluminum electrolysis tests on the inert anodes of this kind of alloy. In the design of the anode shape, factors, such as anode current distribution, oxygen bubble overflow, electrolyte flow, etc., have been fully taken into account, as shown in Fig. 3.43 [159].

It is expected that the development of alloy inert anode materials and the core of a suitable electrolysis process can achieve a dynamic balance between the formation and dissolution of the anode surface protective film during the electrolysis process.

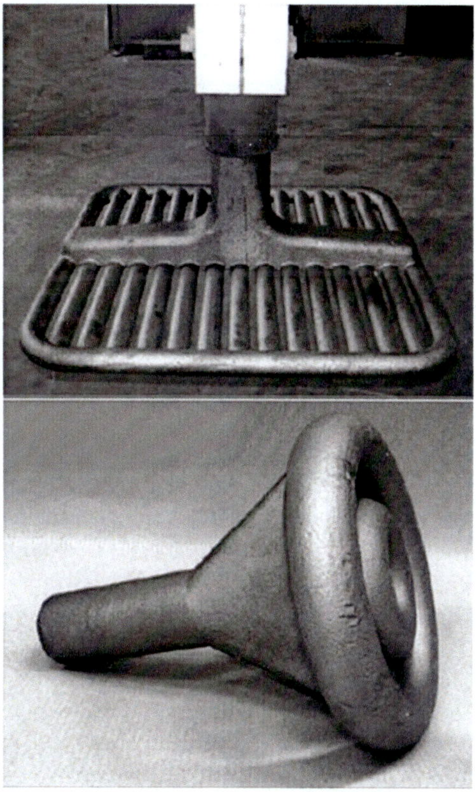

Fig. 3.43 Metal anode prototype developed by Moltech Company [158]

It was necessary to accurately control the formation, dissolution, and diffusion of surface protective film. The surface protective film must be dense and has suitable thickness and good conductivity to protect the metal matrix without hindering the anodic oxygen release reaction. In the electrochemical system of molten salt at high temperature, the formation and dissolution process of the protective film on the anode surface was very complex. In terms of protective film formation, the simple metal in the alloy anode can react with the anode to form new oxygen to form an oxide. This could be transformed into fluoride under appropriate electrochemical conditions, as shown in the equation:

$$[AlOF_2]_2^{2-} + Fe/Ni = 2AlOF + FeF_2/NiF_2 + 2e \tag{3.19}$$

The oxides formed at high-alumina concentration can be transformed into aluminates. The rate of these reactions was closely related to the electrolyte composition, aluminum oxide concentration, temperature, and other electrolytic process parameters, as shown in Fig. 3.44 [160]. In the process of electrolysis, the cell voltage fluctuated when the composition and structure of the protective film changed. When the cell voltage increased, other electrochemical reactions were accelerated with higher decomposition voltage than alumina. For example, the formation of NiF_2 and

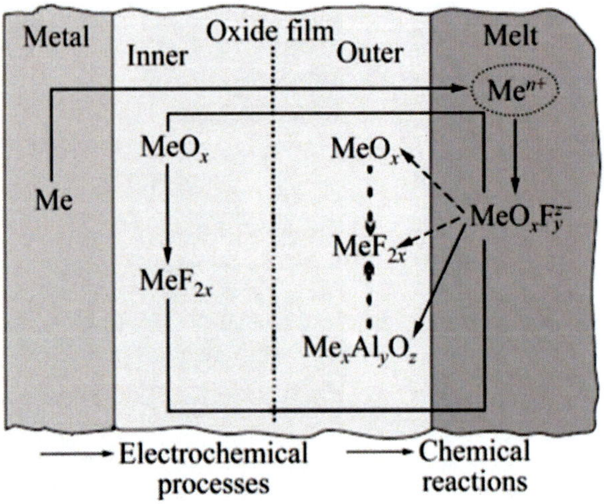

Fig. 3.44 The formation and dissolution of protection film of metallic anode [160]

FeF_2 with poor conductivity was passivated the anode. The protective film on the anode surface cannot grow endlessly, and the thick protective film is not conducive to the electron transmission required for the anodic reaction. Therefore, when the composition and formation rate of the protective film on the anode surface are controlled, the protective film should be dissolved into the electrolyte at the same rate. The ideal thickness of the oxide film should be controlled at 20 μm and 40 μm [161].

In laboratory scale, Helle et al. [162] prepared a series of intermetallic compounds with the general formula $(Cu_{65}Ni_{20}Fe_{15})_{100-x}O_x$ by high-energy ball milling and sintering. They evaluated this composition as an inert anode material for aluminum electrolysis at 700 °C. All compounds had the same composition, and the amount of oxygen added during the grinding process is different. All compounds showed a γ-phase. However, they obtained the best corrosion resistance in low-temperature KF-AlF_3 electrolyte when they added 1.4 wt% of oxygen in the initial alloy.

Nickel-iron alloys had been identified as promising inert anode candidates for the Hall-Heroult process. Chapman, Welch, and Skyllas-Kazacos [163] tested various compositions and then subjected them to short-term galvanostatic electrolysis in a cryolite alumina bath at 960 °C. They oxidized the anodes at 800 °C for 48 h to form a protective scale before electrolysis, in which they identified Fe_2O_3, $Ni_xFe_{3-x}O_4$, and $Ni_xFe_{1-x}O_y$. Anodes having a Ni content of 50–65 wt% performed best during short-time electrolysis. The preformed oxide scale was efficient in reducing anode wear and fluoridation. In the absence of preformed scale, the anodes suffered appreciable internal corrosion and/or passivation due to the metal fluoride formation.

Zhu, He, and Wang [164] prepared Fe-30Ni-5NiO alloys as inert anode materials. The metallic alloy with NiO nanoparticle (mean size around 20 nm) anodes were prepared by spark plasma sintering. The authors investigated the oxidation behavior of these anodes at 800 and 850 °C. A continuous Fe_2O_3 film was formed on the anode surface during electrolysis. Although this alloy exhibits good corrosion resistance to fluoride electrolyte during the electrolysis, this pre-oxidized anode is still not suitable for inert anode applications.

The anodic behavior of surface-oxidized Ni-Fe-Co alloys was investigated over short periods of aluminum electrolysis. Singleton, Welch, and Skyllas-Kazacos [165] found that the addition of 10 wt% Co significantly improve the anodic wear resistance due to the suppression of Fe_xO formation. In general, the protective ability of the preformed oxide scale was greatly affected by the level of porosity and surface adhesion. In electrolytes with less than 4 wt% Al_2O_3, they found that the anodized aluminum had a catastrophic failure due to over polarization at the reaction interface.

Inert anode compositions containing Ni-Fe-Cu were investigated several times. Helle et al. [166] prepared $Cu_xNi_{85-x}Fe_{15}$ ($0 \leq x \leq 85$ wt%) alloys by mechanically alloying. A stable anode for aluminum electrolysis in low-temperature KF-AlF_3 electrolyte was obtained for $65 \leq x \leq 85$. However, a substantial increase of the Cu contamination in produced aluminum was observed for $x > 70$. Goupil et al. [167] evaluated cold spray deposition to produce inert anodes for the aluminum industry. $Cu_{65}Ni_{20}Fe_{15}$ (in wt%) alloy was prepared from the respective metal powders. In the first step, they showed that they could control the particle size of the milled powder with the addition of stearic acid to the initial powder mixture. Then, they optimized the cold spray parameters (pressure and carrier N_2 gas) and produced a thick (1100 μm), dense (porosity 1–2%), and adherent (adhesion strength of 15 MPa) coatings of $Cu_{65}Ni_{20}Fe_{15}$ on nickel aluminum bronze alloy. This allows to envisage the fabrication of large area electrodes required for pilot-scale aluminum electrolysis tests.

Ouvarov-Bancalero et al. studied the influence of adding Yttrium (5 wt%) on the corrosion resistance of the mechanically alloyed $Cu_{65}Ni_{20}Fe_{15}$ [168] anode for aluminum electrolysis in low-temperature KF-AlF_3 electrolyte.

Nguyen [169] optimized an alloy composition containing 64–66% Ni, 25–27% Fe, 7–9% Mn, 0–0.7% Cu, and 0.4–0.6% Si. The weight ratio of Ni/Fe is preferably 2.3–2.6, the weight ratio Ni/Ni + Cu is greater than 0.98, the weight ratio Cu/Ni is less than 0.01, and the weight ratio Mn/Ni is 0.09–0.15. The alloy surface includes $NiFe_2O_4$ produced by pre-oxidation of the alloy.

3.11 Engineering Tests

3.11.1 Cermet Inert Anode

American Aluminum Company and Reynolds Metals Company carried out a 25-day test of 6 kA grade cermet inertia anode electrolysis with the support of the US Department of Energy [170]. A single research team led by Chinalco and Zhongnan University of China carried out a 28-day electrolytic test on 4 kA cermet inertia anode. It is said that the experiment will be carried out on a larger scale and for a longer period of time. However, there is no further report [171].

Li et al. [172] conducted the simulation of heat balance in 5 kA grade aluminum reduction cell with inert anodes.

The cell profile and heat balance of 5 kA grade aluminum reduction cell with inert anodes were calculated and simulated by finite element software Ansys. The results showed that the superheat temperature and the type of side carbon block had a great influence on the cell profile. When the superheat temperature increases by 10 °C, the ledge thickness was reduced by 50%, and the ledge width was reduced by 30%. The better the heat dissipation capacity of the side carbon block, the thicker the battery wall shelf. It is recommended to use a semi-graphitic side carbon block because it not only assured the required cell profile but also reduced the heat dissipation. The thermal balance calculations showed that both the cell with semi-graphitic cathode and graphitized cathode could achieve thermal balance by taking some measures and the cell with graphitized cathode needed about 9% more energy input (Figs. 3.45, 3.46, 3.47, 3.48 and 3.49).

Biao et al. [173] conducted the pilot test of aluminum electrolysis by the $NiFe_2O_4$-M inert anode.

A pilot test of an aluminum electrolysis was completed in 2014 using inert anodes with the composition $NiFe_2O_4$-M (M is metal). The volume of the cell was 110 × 66 × 79 cm, and the size of inert anode was Φ34 × 22 cm. The total number of inert anodes in the cell were 6. The current density of the inert anode was about

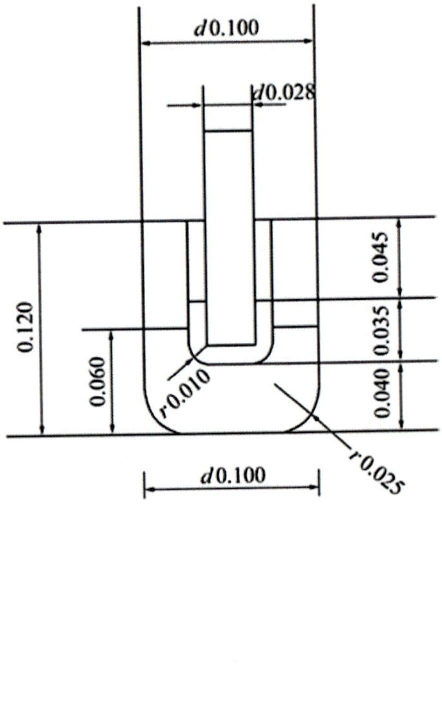

Fig. 3.45 The section dimension for inert anode (m)

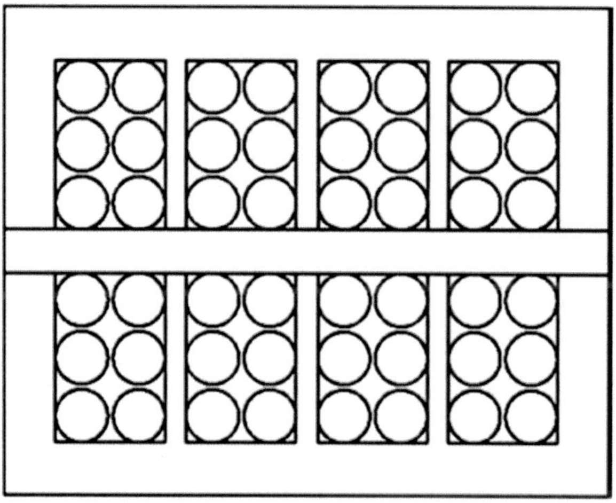

Fig. 3.46 The anode configuration in cell

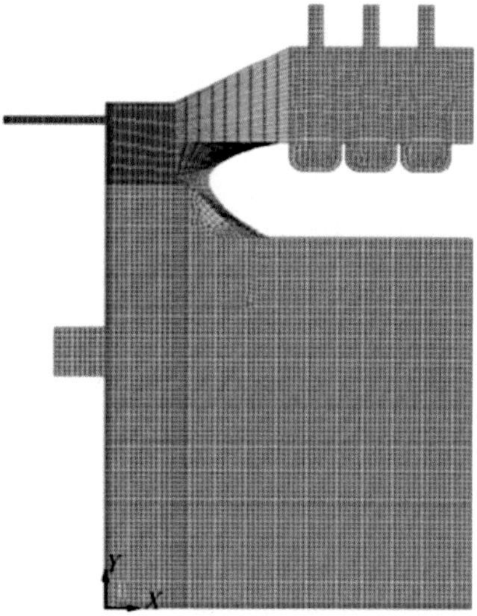

Fig. 3.47 Cell profile of calculation model

0.8 A/cm^2, the total electrolysis current fluctuated from 2700 to 4000 A, and the electrolysis duration was 500 h. The pilot test results show that the electrolysis process can be successfully carried out due to the good conductivity and corrosion resistance of the abovementioned inert anode. In addition, none of the six inert anodes cracked during the whole electrolysis process even at the beginning of electrolysis when they were inserted in high-temperature electrolyte. The main problem of pilot test was that the content of Fe as impurity was about 1.7% in the product,

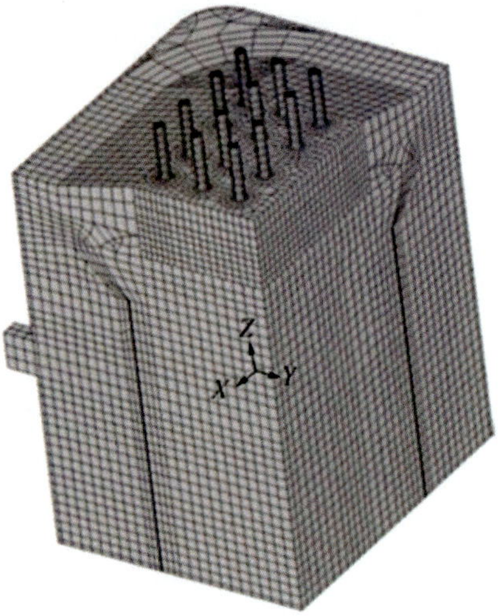

Fig. 3.48 The thermal balance calculation model

which was higher than the baseline value of GB/T. In order to remove Fe, related experiments were continually conducting (Figs. 3.50, 3.51, 3.52, and 3.53 and (Tables 3.15, 3.16 and 3.17).

3.11.1.1 Results

1. After sintering, the size of the $NiFe_2O_4$-M cermet inert anode was significantly reduced, and the density is about 5.391 g/cm^3. The strength and toughness were significantly improved. In addition, the inert anodes did not break after being dropped on cement floor from a height of 1.5 m, and the inert anodes may withstand cutting work.
2. The sintered $NiFe_2O_4$-M cermet inert anode performed well in the pilot test aluminum electrolysis. These inert anodes did not crack during the heat rush. This shows excellent electrical conductivity and corrosion resistance throughout the electrolysis process.
3. The impurity of Fe content in electrolytic aluminum product exceeded the baseline level of GB/T Al 99.00, and Fe mainly came from Fe_2O_3 component of cermet inert anode. In order to overcome this problem, complementary test continued.

The electrolysis technology of Electrolysis Co., Ltd., co-founded by American Aluminum Company and Rio Tinto, was the most representative. It was inferred that most of the problems with the inert anode and the inert cathode material had been solved. It was likely to be a diversion cell and to be done by quasi-continuous feeding method with 0.5 A/cm^2 or lower current density. An inert anode based on

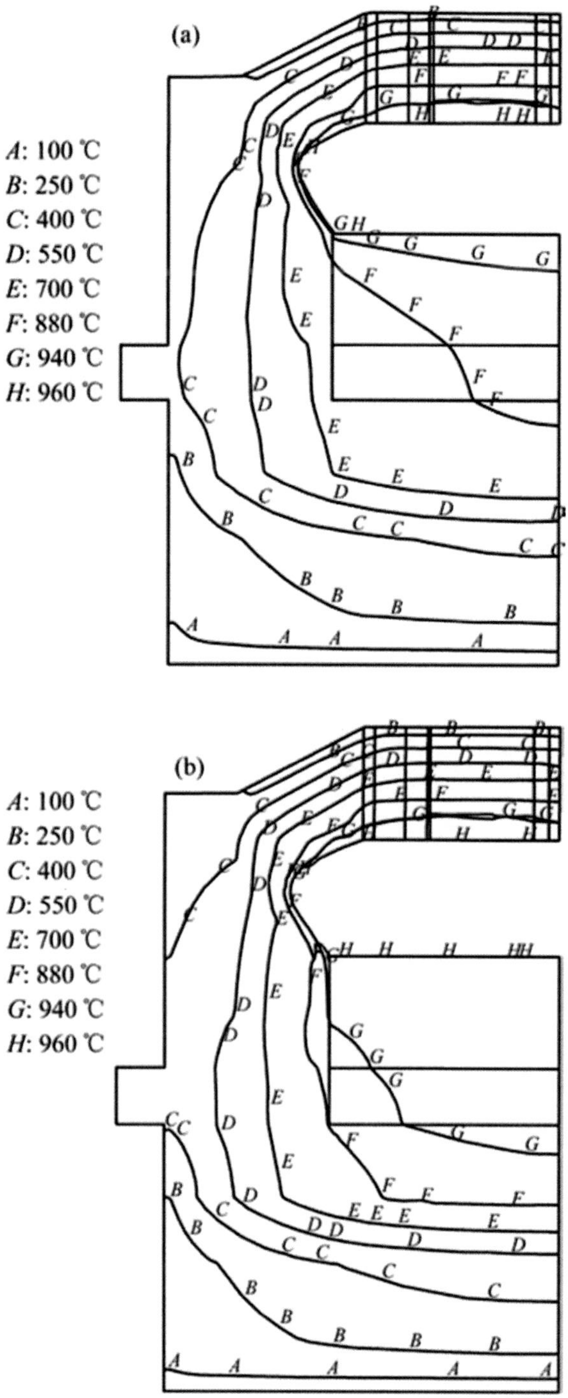

Fig. 3.49 Temperature distribution comparison of cell section. (**a**) Common cathode; (**b**) graphitized cathode

Fig. 3.50 A pair of $NiFe_2O_4$-M cermet inert anodes were connected together

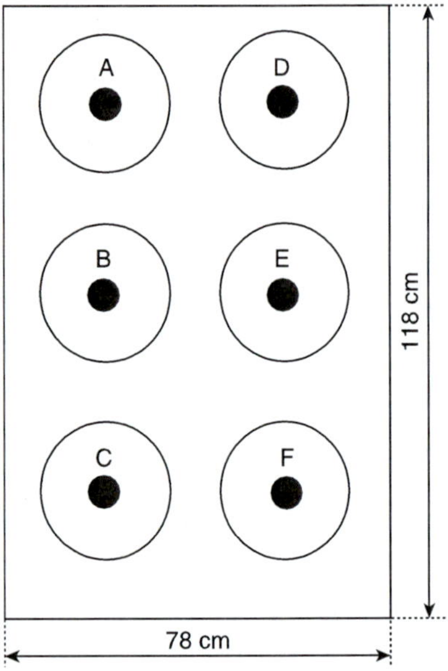

Fig. 3.51 Overhead view of $NiFe_2O_4$-M cermet inert anodes distribution in the cell

Fig. 3.52 The picture of NiFeO-M cermet inert anode after sintering

Fig. 3.53 The inert anodes distribution in the cell. The picture of $NiFe_2O_4$-M cermet inert anode after 100 h electrolysis (on a small scale)

Table 3.15 Time distribution at different stages in pilot test

No.	Time (h)	Percent (%)
1. Roast and starting of the cell	120	24
2. Replacing anodes	8	1.6
3. Electrolysis by cermet inert anodes	368	73.6
4. End of cell	4	0.8

Table 3.16 Sizes of $NiFe_2O_4$-M cermet inert anode measured before and after 100 h electrolysis

	Diameter (mm)	Height (mm)
Before electrolysis	25.54	67.54
After 100 h electrolysis	25.48	67.47

Table 3.17 The contents of impurity in electrolyzed Al and concerned standards

Elements	Pilot test results (%)	Baseline level GB/T (%)
Al	97.4811	99.00
Fe	1.3317	0.50
Si	0.9090	0.42
Ti	0.1293 -	
Zn	0.0250	0.05
Ga	0.0249	0.05
Cu	0.0240	0.05
Mn	0.0199 -	
Ni	0.0189-	

metal ceramics was used, and the replacement period may be between 2 and 3 years, or it may pass through the overhaul of the stop slot and replace it as a whole [174].

3.11.2 Alloy Anode

The alloy anode was mainly from Rusal (UC Rusal) and Swedish Moltech Company.

The Ni-Fe alloy anode of Moltech Company in Sweden had attracted much attention. 4 kA and 25 kA aluminum electrolysis experiments were carried out on this kind of alloy anode. The anode current distribution, oxygen bubble overflow, electrolyte flow, and other factors have been fully taken into account in the design of anode shape.

After the dissolution of Moltech Company, some technicians transferred to Rusal (UC Rusal), and Rusal also had their own inert anode research and development foundation [174, 175].

Simakov and co-workers [176] reported that approximately 270 h of laboratory electrolysis tests were conducted at the Rusal Engineering and Technology Center in Krasnoyarsk. Tests showed that the consumption of inert anodes is less than 2.5 cm/year and the aluminum purity is better than 99.2%. In the near future, inert anode tests of more than 2 kA will be developed, and then batteries that can work at currents of more than 100 kA will be developed. The anode metal structure contained two phases due to the separation of a liquid solution into two solid solutions: a solid solution of copper in iron and a solid solution of iron in copper. Nickel was dissolved in both phases.

In 2013, UC Rusal a Russian aluminum joint company, announced that it had begun a bench test of industrial smelting electrolyzers based on inert anode technology. The bench test at Krasnoyarsk plant will be stopped when the current intensity reaches 3 KVA. The company developed a new set of electrolytic cell separation test scheme for inert anode.

The anode technology Rusal United plans to test the production of Kraz's inertia anode electrolyzer at Krasnoyarsk aluminum smelter in 2015, after a successful bench test. The company was likely to transform its smelter to inert anode technology from 2017, started with the Krasnoyarsk plant [177].

In 2015, Rusal (UC Rusal) began to do the industrial test (100 kA) of the inert anode aluminum reduction cell. The alloy inert anode (similar to the alloy anode of Moltech Company) was used, with oxide film on the surface and horizontal arrangement. The results of the initial industrial test cell of Rusal (UC Rusal) were not particularly good, and the power consumption was higher, which was 3000° higher than that of carbon anode electrolyzer per ton of aluminum. However, there was no formal public report [174].

In 2019, Viktor Mann [178] reported that Rusal was implementing an environmental strategy that includes increasing the number of prebaked (PB) pot lines and implementing global environmental standards at the existing smelters. The following result has been achieved, including the construction of a testing area for inert anode cells that are intended to replace Vertical Stud Soderberg (VSS) cells. This indicates that Rusal industry may have achieved considerable success in the inert anode technology.

3.11.3 Environmental Protection

Viktória Barbara Kovács et al. [179] conducted a comparative analysis on the environmental impact of aluminum smelting technology. This paper focused on the global warming potential (GWP) of primary aluminum production and analyzed the hot spots of greenhouse gas (GHG) emission. The major GHG emission of primary aluminum production is caused by aluminum smelting. Inert anodes or non-consumable anodes were considered as a solution to mitigate GHG emission of the Hall-Heroult process.

A comparative cradle-to-gate life cycle assessment (LCA) was conducted to estimate the environmental impact of primary aluminum production considering conventional and inert anode smelting. This includes the production technology of the anodes and the influence of the energy mix used to produce the electrical power.

Asbjørn Solheim et al. [180] reported that the introduction of inert anodes in the alumina reduction tank may bring some advantages. However, it also brings many serious defects in the blind lane research center where inert anodes are environmentally friendly. In view of the developments in the Hall-Heroult process over the past few decades, it was desirable to conduct a rigorous evaluation of the inert anode concept compared to state-of-the-art electrolysis cells. With an inert anode, the DC energy consumption will increase by 3 kWh/kg-Al, partly because the 1 V higher isothermal cell voltage cannot be fully compensated. This is mainly because a cell with inert anodes requires similar thermal loss as a cell with carbon anodes. Therefore, the total carbon dioxide footprint will be higher with inert anodes when the power is generated in a coal-fired plant. In addition, if the power comes from a gas-fired plant, there is no significant difference. The full carbon dioxide reduction potential with inert anodes can only be realized when using renewable energy. However, a Hall-Heroult plant with carbon capture and sequestration will consume less electrical power than a plant with inert anodes.

Carbon-free reduction technology is a technology with inert anode as its core. In the next 10 years, the possibility of industrialization of inert anode is very high. However, the problems of short anode life, high thermal loss, and high power consumption of cell need to be solved. In the aspect of environmental protection, if coal-fired power is used, the carbon emission of inert anode aluminum reduction is higher than that of carbon anode. If clean energy such as hydropower is used, the carbon emission of inert anode aluminum reduction will be lower than that of carbon anode. The main advantages of inert anode aluminum reduction are that they completely eliminate the harm of coal tar pitch and improve the environment of the electrolytic workshop. It may have certain advantages and competitiveness in terms of yield and comprehensive cost [174].

References

1. De Nora, V.: Veronica and Tinor 2000 new technologies for aluminum production. Electrochem. Soc. Interface. **11**(4), 20–24 (2002)
2. Nancy, M., Jack, E.: Inert anode roadmap: A framework for technology development, pp. 1–29. The Aluminum Association, Washington, DC (1998)
3. Noel, J.: Future developments in the Bayer-Hall-Héroult process. In: Burkin, A.R. (ed.) Production of aluminium and alumina, pp. 188–207. Wiley, Chichester (1987)
4. Qiu, Z.X.: Prebaked anode electrolysis cell aluminium smelting, p. 488. Metallurgical Industry Press, Beijing (2005)
5. Pawlek, R.P.: Modern inert anode. In: Light Metals 2002, pp. 449–456. TMS, Warrendale, PA (2002)
6. Duruz, J.J., de Nora, V.: Aluminium electrowinning cells with Ni-Fe alloy anodes, W.O. Patent 01/43208, (1999)
7. Altdorfer, T.: DOE follows roadmap to aluminium technologies of the future. JOM. **52**(11), 19–25 (2000)
8. van Leuwen, T.M.: An aluminum revolution, p. 29. Credit Suisse First Boston, New York (2000)
9. Sadoway, D.R.: Inert anodes for the Hall-Héroult cell: The ultimate materials challenge. JOM. **53**(5), 34–35 (2001)
10. Keniry, J.: The economics of inert anodes and wettable cathodes for aluminum reduction cells. JOM. **53**(5), 43–47 (2001)
11. Hall, C.M.: Process of electrolyzing fused salts of aluminum. US Patent 400, 667. Filed September 21, 1888. Patented April 2, 1889
12. Billehaug, K., Oye, H.A.: Inert anodes for aluminium electrolysis in Hall-Héroult cells (I). Aluminium. **57**(2), 146–150 (1981)
13. Billehaug, K., Oye, H.A.: Inert anodes for aluminium electrolysis in Hall-Heroult cells (H). Aluminium. **57**(3), 228–231 (1981)
14. McLeod, A.N., Hagerty, J.S., Sadoway, D.R.: Inert anode materials for Hall cells. In: Light Metals 1986, pp. 269–273. TMS, Warrendale, PA (1986)
15. Ray, S.: Inert anodes for Hall cells. In: Light Metals 1986, pp. 287–298. TMS, Warrendale, PA (1986)

16. DeYoung, D.H.: Solubility of oxides for inert anodes in cryolite-based melts. In: Light Metals 1986, pp. 299–307. TMS, Warrendale, PA (1986)
17. Tarcy Gary, P.: Corrosion and passitivity of cermet inert anodes in cryolite-type electrolytes. In: Light Metals 1986, pp. 309–320. TMS, Warrendale, PA (1986)
18. Grjotheim, K., Krohn, C., Malinovsky, M., et al.: Aluminium electrolysis-Fundamentals of the Hall-Heroult process, 2nd edn, p. 365. Aluminium-Verlag, Dusseldorf (1982)
19. Thonstad, J., Fellner, P., Haarberg, G.M., et al.: Aluminium electrolysis-Fundamentals of the Hall-Heroult process, 3rd edn, p. 279. Aluminium-Verlag, Dusseldorf (2001)
20. Horinouchi, K., Tachikawa, N., Yamada, F.C.: DSA in aluminum reduction cells. In: Organizing Committee of the First International Symposium on Molten Salt and Technology (ed.) Proceedings of the first international symposium on molten salt chemistry and technology, p. 65. Molten Salt Committee of the Electrochemical Society of Japan, Kyoto, Japan (1983)
21. Keller, R., Rolseth, S., Thonstad, J.: Mass transport considerations for the development of oxygen-evolving anodes in aluminium electrolysis. Electrochim. Acta. **42**(12), 1809–1917 (1997)
22. Augustin CO, Srinivasan LK, Srinivasan KS. Inert anodes for environmentally clean production of aluminium. Bull. Electrochem., 1993, 9(8–10): 502–503
23. Xianjin, Y., Zhuxian, Q., Songzhe, J.: Corrosion of zinc ferrite in NaF-AlF3-Al2O3 molten salts. J. Chin. Soc. Corrosion Protect. **20**(5), 275–279 (2000)
24. Galasiu, R., et al.: Inert anodes for aluminium electrolysis: Variation of the properties of nickel ferrite ceramics as a function of the way of preparation. In: Haarberg, G.M. (ed.) Eleventh Intemational Aluminium Symposium. Norway, September 19–22, pp. 133–136 (2001)
25. Zhang, Y., Wu, X., Rapp, R.A.: Modeling of the solubility of NiO/NiAl2O4 and Fe0/FeAl2O4 in cryolite melts. In: Crepeau, P.N. (ed.) Light Metals 2003, pp. 415–421. TMS, Warrendale, PA (2003)
26. Pawlek, R.P.: Inert anodes: An update. In: Tabereaux, A.T. (ed.) Light Metals 2003, pp. 283–287. TMS, Warrendale, PA (2004)
27. Yang, J.H., Liu, Y.X., Wang, H.Z.: The behavior and improvement of SnO2 based anodes in aluminum electrolysis. In: Mason, D.A. (ed.) Light Metals 1993, pp. 493–495. TMS, Warrendale, PA (1993)
28. Qiu, Z.X., Fan, L.M.: The rate-determining step of metal loss in cryolite alumina melts. In: McGeer, J.P. (ed.) Light Metals 1984, pp. 789–804. TMS, Warrendale, PA (1984)
29. Haarberg, G.M.: The interaction between tin oxide and cryolite-alumina melts. In: Ninth international symposium on molten salts, 22–27 May, pp. 568–577. Electrochemical Society, San Francisco, CA (1994)
30. Issaeva, L.: Electrochemical behaviour of tin dissolved in cryolite-alumina melts. Electrochim. Acta. **42**(6), 1011–1018 (1997)
31. Yang, J.H., Thonstad, J.: On the behaviour of tin-containing species in cryolite-alumina melts. J. Appl. Electrochem. **27**, 422–427 (1997)
32. Sadus, A.M.V., Constable, D.C., Dorin, R., et al.: Tin dioxide-based ceramics as inert anodes for aluminum smeltering: A laboratory study. In: Hale, W. (ed.) Light Metals 1996, pp. 259–265. TMS, Warrendale, PA (1996)
33. Popescu, A.M., Constantin, V.: The dependence of current efficiency on the operating parameters in aluminium electrolysis cell with SnO2-based inert anodes. Rev. Roum. Chim. **43**(9), 793–798 (1998)
34. Galasiu, R.: SnO2-based inert anodes for aluminium electrolysis. Influence of Ag20 on the electrical and electrochemical properties. In: X. Al Symposium, Slovak-Norwegian Symposium on Aluminium Smelting Technology, Stara Lesna-Ziar nad Hronom, 21–23 Sept. 1999, pp. 35–38 (1999)
35. Las, W.C.: Influence of additives on the electrical properties of dense SnO2-based ceramics. J. Appl. Phys. **74**(10), 6191–6196 (1993)

36. Cassyre, L., Utigard, T.A., Bouvet, S.: Visualizing gas evolution on graphite and oxygen evolving anodes. JOM. **54**(5), 140–149 (2000)
37. Duruz, J.J, Derivaz, J.P., Debely, P.E., et al.: Molten salt electrowinning method, anode and manufacture thereof. US Patent 4614569, (1986)
38. Walker, J.K., Kinkoph, J., Saha, C.K.J.: Appl. Electrochem. **19**, 225–230 (1989)
39. Dewing, E.W., Reesor, D.N.: US Patent, 4668351 (1987)
40. Gregg, J.S., Frederick, M.S., King, H.L., et al.: Testing of cerium oxide coated cermet anodes. In: Subodh, K.L.D. (ed.) Light Metals 1993, pp. 455–464. TMS, Warrendale, PA (1993)
41. Dewing, E.W., et al.: The chemistry of solutions of CeO2 in cryolite melts. Metallur. Mater. Trans. B. **26B**(I), 81–86 (1995)
42. Yexiang, L.: Functional electrode material and its application, p. 142. Central South University of Technology Press, Changsha (1996)
43. Pietrzyk, S., Oblakowsky, R.: Concentration of impurities from the inert anodes in the bath and metal during aluminium electrolysis. In: X. AJ Symposium, Slovak-Norwegian, Symposium on Aluminium Smelting Technology, Stara Lesna-Ziar nad Hronom, 21–23 Sept, pp. 31–34 (1999)
44. Zaikov, Y.P., et al.: Ceramic properties of electrodes based on NiO-Li2O and their solubility in cryolite alumina melts. In: W. Al Symposium, 25–27 Sept, Slovakia, Ziar nad Hronom-Donovaly, pp. 239–241 (1995)
45. El Sadoway, D.: Inert anodes for the Hail-Héroult cell: the ultimate materials challenge. JOM. **53**(5), 34–35 (2001)
46. Hryn, J.N., Pellin, M.J.: A dynamic metal anode. In: Eckert, C.E. (ed.) Light Metals 1999, pp. 377–381. TMS, Warrendale, PA (1999)
47. Yang, J.H., Hryn, J.N., Davis, B.E., et al.: New opportunities for aluminum electrolysis with metal anodes in a low temperature electrolyte system. In: Tabereaux, A.T. (ed.) Light Metals 2004, pp. 321–326. TMS, Warrendale, PA (2004)
48. Yang, J.H., Hryn, J.N., Krumdic, G.K.: Aluminum electrolysis tests with inert anodes in KF-A1F3-based electrolytes. In: Galloway, T.J. (ed.) Light Metals 2006, pp. 421–424. TMS, Warrendale, PA (2006)
49. Beck, T.R.: Production of aluminum with low temperature fluoride melts. In: Mannweiler, U. (ed.) Light Metals 1994, pp. 417–423. TMS, Warrendale, PA (1994)
50. Sekhar, J.A., Deng, H., Liu, J., et al.: Micropyretically synthesized porous non-consumable anodes in the Ni-Al-Cu-Fe-X system. In: Huglen, R. (ed.) Light Metals 1997, pp. 347–354. TMS, Warrendale, PA (1997)
51. Duruz, J.J., de Nora, V.: Aluminium electrowinning with metal-based anodes. WO 0142535, 2001-06-14
52. Bradford, D.R.: Inert anode metal life in low temperature reduction process final technical report, p. 65. Goldendale Aluminum Company (2005)
53. Thinh, N., Vittorio, B., Curtis, M., et al.: Non-carbon anodes and cathode coalings for aluminum production. JOM. **56**(11), 231–237 (2004)
54. Nguyen, T., de Nora, V.: Oxygen evolving inert metallic anode. In: Calloway, T.J. (ed.) Light Metals 2006, pp. 385–390. TMS, Warrendale, PA (2006)
55. De Nora, V., Nguyen, T., Von Kaenel, R., et al.: Semi-vertical de NORA inert metallic anode. In: Sorlie, M. (ed.) Light Metals 2007, pp. 501–505. TMS, Warrendale, PA (2007)
56. Antilie Klinger, L., Von Kaenel, R., et al.: Modeling of a 25 kA de NORA inert metallic anode test cell. In: Galloway, T.J. (ed.) Light Metals 2006, pp. 391–396. TMS, Warrendale, PA (2006)
57. Von Kaenel, R., de Nora, V.: Technical and economical evaluation of the de Nora inert metallic anode in aluminum reduction cells. In: Galloway, T.J. (ed.) Light Metals 2006, pp. 397–402. TMS, Warrendale, PA (2006)
58. Weyand, J.D., DeYoung, D.H., Ray, S.P., et al.: Inert Anodes for Aluminium Smelting. Aluminum Company of America, Washington, DC (1986)

59. Ray, S.P.: Inert anode for Hall cells. In: Miller, R.E. (ed.) Light Metals 1986, pp. 287–298. TMS, Warrendale, PA (1986)
60. Alcom, T.R., Tabereaux, A.T., Richards, N.E., et al.: Operational results of pilot cell test with cermet inert anodes. In: Das, S.K. (ed.) Light Metals 1993, pp. 433–443. TMS, Warrendale, PA (1993)
61. Windusch, C.F., Strachan, D.M., Henager, C.H.: Material characterization of cermet anodes tested in a pilot cell. In: Das, S.K. (ed.) Light Metals 1993, pp. 445–454. TMS, Warrendale, PA (1993)
62. Gregg, J.S., Frederick, M.S., King, H.L., et al.: Testing of cerium oxide coated cermet anodes in aluminum laboratory cell. In: Das, S.K. (ed.) Light Metals 1993, pp. 455–464. TMS, Warrendale, PA (1993)
63. Gregg, J.S., Frederick, M.S., Vaccaro, A.J.: Pilot cell demonstration of cerium oxide coated anodes. In: Das, S.K. (ed.) Light Metals 1993, pp. 465–473. TMS, Warrendale, PA (1993)
64. Kozarek, K.I., Ray, S.P., Dawless, R.K., et al.: Corrosion of cermet anodes during low temperature. In: Electrolysis of alumina. DE-FC07-89ID12848. Alcoa, Pittsburgh, PA (1997)
65. Kozarek, K., Ray, S.P., Dawless, R.K., et al.: Energy efficient aluminum production—pilot-scale cell tests: Final report for phase I and phase E. DE-FC07-941D132I9. Alcoa, Pittsburgh, PA (1999)
66. Christini, R.A., Dawless, R.K., Ray, S.P., et al.: Phase IH advanced anodes and cathodes utilized in energy efficient aluminum production cells. DE-FC07-98ID13666. Alcoa, Pittsburgh, PA (2001)
67. Ray, SP.: Electrolytic production of high purity aluminium using ceramic inert anodes, US Patent, 6,416,649 (2001)
68. Blinov, V., Polyakov, P.: Behaviour of cermet inert anodes for aluminium electrolysis in a low temperature electrolyte. In: Haarberg, G.M. (ed.) 11th International Aluminium Symposium. Norway, September 19–22, pp. 123–131 (2001)
69. Lorentsen, O.A., Thonsiadl, J.: Laboratory cell design considerations and behaviour of inert anodes in cryolite-alumina melts. In: Haarberg, G.M. (ed.) 11th International aluminium symposium. Norway, September 19–22, pp. 145–154 (2001)
70. Blinov, V., Polyakov, P., Krasnoyarsk, et al.: Behaviour of inert anodes for aluminium electrolysis in a low temperature electrolyte, part I. Aluminium. **73**(12), 906–910 (1997)
71. Dawless, R.K., et al.: Reduced temperature aluminium production in an electrolytic cell having inert anodes. US patent, 6,030,518 (1997)
72. Office of Industrial Technologies: Advanced Anodes and Cathodes Utilized in Energy-Efficient Aluminium Production Cells, p. 2. Aluminum Project Fact Sheet (1999)
73. Belda, A.: Alcoa quarterly financial report to analysts (2001)
74. Baker Frank, W., Rolf Richard, L.: Hall cell with inert anode. In: Light Metals 1986, pp. 275–283 (1986)
75. DeYoung, D.H.: Solubilities of oxides for inert anode in cryolite-based melts. In: Miller, R.E. (ed.) Light Metals 1986, pp. 299–307. TMS, Warrendale, PA (1986)
76. Olsen, E., Thonstad, J.: The behaviour of nickel ferrite cermet materials as inert anodes. In: Hale, W. (ed.) Light Metals 1996, pp. 249–257. TMS, Warrendale, PA (1996)
77. Tarcy, G.P.: Corrosion and passivation of cermet inert anodes in cryolite-type electrolytes. In: Miller, R.E. (ed.) Light Metals 1986, pp. 309–320. TMS, Warrendale, PA (1986)
78. Windisch, C.F., Marschman, S.C.: Electrochemical polarization studies on Cu and Cu ~ containing cermet anodes for the aluminum industry. In: Zabreznik, R.D. (ed.) Light Metals 1987, pp. 351–355. TMS, Warrendale, PA (1987)
79. Olsen, E., Thonstad, J.: Nickel ferrite as inert anodes in aluminum electrolysis: Part D. Material performance and long-term testing. J. Appl. Electrochem. **29**(3), 301–311 (1999)
80. Lorentsen, O.A., Thonstad, J.: Electrolysis and post-testing of inert cermet anodes. In: Schneider, W. (ed.) Light Metals 2002, pp. 457–462. TMS, Warrendale, PA (2002)
81. Xinzheng, L.: Study on Sintering, Conductivity and Corrosion Resistance of xM/(10NiO-$NiFe_2O_4$) Ceramic [Master Thesis]. Central South University, Changsha (1992)

82. Xiao, H.M.: On the corrosion and the behavior of inert anodes in aluminium electrolysis: [Doctor Thesis]. Norwegian Institute of Technology, Trondheim, Norway (1993)
83. Jianhong, Y.: Simulation study of inert electrode and bipolar multichamber cell in aluminum electrolysis [Doctoral thesis]. Central South University of Technology, Changsha (1992)
84. Diep, Q.B., Dewing, E.W., Sterten, A.: The solubility of Fe2O3 in cryolite-alumna melts. Metallur. Mater. Trans. B. **33B**(1), 140–142 (2002)
85. Wang, H.Z., Thonstad, J.: The behaviour of inert anodes as a function of some operating parameters. In: Campbell, P.G. (ed.) Light Metals 1989, pp. 283–290. TMS, Warrendale, PA (1989)
86. Cao, X.Z., Wang, Z.W., Shi, Z.N., et al.: Al-Ti~O-X cermet as inert anode for aluminum electrolysis. In: Sorlie, M. (ed.) Light Metals 2007, pp. 927–929. TMS, Warrendale, PA (2007)
87. Lorentsen, H.J.: Inert anode under electrochemical impedance spectroscopy study. In: Haarberg, G.M. (ed.) Eleventh International Aluminium Symposium. Norway, September 19–22, pp. 137–143 (2001)
88. Yankeelov, J.: Inert anode metal life in low temperature aluminum reduction process. DE-FC07-98ID13662. Northwest Aluminum Technologies, Dalles, OR (2003)
89. Olsen, E., Thonstad, J.: Nickel ferrite as inert anodes in aluminum electrolysis: Part I. Materials fabrication and preliminary testing. J. Appl. Electrochem. **29**(3), 293–299 (1999)
90. John Berchmans, L., et al.: Evaluation of Mg2+−substituted NiFeO4 as a green anode material. Mater. Lett. **58**(5), 1928–1933 (2004)
91. Wanli, J., Lei, Z., Guangchun, Y., Yihan, L.: Sintering process of $NiFe_2O_4$ spinel with and without TiO_2 adding. J. Chin. Ceram. Soc. **32**(9), 1150–1153 (2004)
92. Jinhui, X., Guangchun, Y., Yihan, L., Xiaoming, Z.: Effects of additive V_2O_5 on sintering mechanism and properties of nickel ferrite. J. Chin. Ceram. Soc. **33**(6), 683–687 (2005)
93. Jinhui, X., Guangchun, Y., Yihan, L.: Effect of additives on microstructure and properties of inert anode of $NiFe_2O_4$ spinel. J. Northeastern Univ. (Nat. Sci.). **26**(6), 575–577 (2005)
94. Ray, S.P., Woods, R.W., Dawless, R.K., et al.: Electrolysis with an Inert Electrode Containing Ferrite, Copper and Silver, US Patent, 5,865,980, April 26, 1997
95. Yanqing, L., Jie, L., Zhongliang, T., et al.: An improved pyroconductivity test of spinel-containing cermet inert anodes in aluminum electrolysis cells. In: Tabereaux, A.T. (ed.) Light Metals 2004, pp. 339–344. TMS, Warrendale, PA (2004)
96. Strachan, D.M., Koski, O.H., Morgan, L.G., et al.: Results from a 100-hour electrolysis test of a cermet anode: materials aspects. In: Bickert, C.M. (ed.) Light Metals 1990, pp. 395–401. TMS, Warrendale, PA (1990)
97. Peterson, R.D., Richards, N.E., Tabereaux, A.T.: Results of 100-hour electrolysis test of a cermet anode; operational results and industry perspective. In: Bickert, C.M. (ed.) Light Metals 1990, pp. 385–393. TMS, Warrendale, PA (1990)
98. De Nora, V., Duruz, J.J.: Aluminum electrowinning operating with Ni-Fe alloy anodes. WO patent 01/43, 208, Dec. 9, 1999
99. Thonstad, J., Rolseth, S.: Alternative electrolyte compositions for aluminium electrolysis. Transactions of the Institutions of Mining and Metallurgy, Section C. Mineral Processing and Extractive Metallurgy. **114**(3), 188–191 (2005)
100. Beck, T.R.: A nonconsumable metal anode for production of aluminum with low temperature fluoride melts. In: Light Metals 1995, pp. 355–360 (1995)
101. Brown, C.W.: Next generation vertical electrode cells. JOM. **53**(5), 39–42 (2001)
102. Brown, C.W.: Laboratory experiments with low-temperature slurry-electrolyte alumina reduction cells. In: Peterson, R.D. (ed.) Light Metals 2000, pp. 391–396. TMS, Warrendale, PA (2000)
103. De Nora, V., Duruz, J.J.: Low temperature operating ceil for the electrowinning of aluminum, WO patent 01/31, 086, Oct 26, 1999
104. Beck, T.R., Brown, C.W.: Aluminum low temperature smelting cell metal collection. US Patent 6,419,812, Jul 16, 2002

105. Brown, C.W., Frizzle, P.B.: Low temperature aluminum reduction cell using hollow cathode. US Patent 6,436,272, Aug. 20, 2002
106. Wang, J.W., Lai, Y.Q., Tian, Z.I., et al.: Investigation of 5Cu-(10NiO-NiFe2O4) inert anode in low-temperature aluminium electrolysis. In: Sorlie, M. (ed.) Light Metals 2007, pp. 525–530. TMS, Warrendale, PA (2007)
107. Kryukovsky, V.A., Frolov, A.V.T., Tkatcheva, O.Y., et al.: Electrical conductivity of low melting cryolite melts. In: Galloway, T.J. (ed.) Light Metals 2006, pp. 409–413. TMS, Warrendale, PA (2006)
108. Thonstad, J., Fellner, P., Haarberg, G.M., et al.: Aluminium electrolysis-fundamentals of the Hall-Héroult process, 3rd edn, p. 10. Aluminium-Verlag, Dusseldorf (2001)
109. Oye, H.A., Mason, N., Peterson, R.D., et al.: Aluminum: Approaching the new millennium. JOM. **51**(2), 29–42 (1999)
110. Yexiang, L.: Progress in research and development of inert anode and wettable cathode for aluminum electrolysis. In: Light Metals 2001, vol. 5, pp. 26–29 (2001)
111. Sorlie, M., Oye, H.A.: Cathodes in aluminum electrolysis, 2nd edn, pp. 66–73. Aluminium-Verlag, Dusseldorf (1994)
112. McMinn, C.J.: A review of RHM cathode development. In: Cutshall, E.R. (ed.) Light Metals 1992, pp. 419–425. TMS, Warrendale, PA (1992)
113. Pawlek, R.P.: Cathodes wettable by molten aluminium for aluminum electrolysis cells. Aluminium. **66**(8), 573–582 (1990)
114. Qingyu, L.: Preparation and Industrial Application of Inert Wettable TiB2 Composite Cathode Coating for Aluminum Electrolysis [Doctoral Thesis]. Central South University, Changsha (2003)
115. Richards, N.E.: Electrolytically conductive cermet compositions. US Patent, 3,328,280
116. Kaplan, H.I.: Cathodes for alumina reduction cells. US Patent, 4,333,813, 1982-6-8
117. Juel, L.H., Joo, L.A., Tucker, K.W.: Composite of TiB2-graphite. US Patent, 4,465,581. 1984-8-14
118. Tucker, K.W., Gee, J.T., Shaner, J.R., et al.: Stable TiB2 graphite cathodes for aluminum production. In: Welch, B.J. (ed.) Light Metals 1987, pp. 345–349. TMS, Warrendale, PA (1987)
119. Alcom, T.R., Stewart, D.V., Tdbereaux, A.T., et al.: Pilot reduction cell operation using TiB2~G cathodes. In: Blckert, C.M. (ed.) Light Metals 1990, pp. 413–418. TMS, Warrendale PA (1990)
120. Tabereaux, A., Brown, J., Eldridge, I., et al.: The operational performance of 70kA prebake cells retrofitted with TiB2~G cathode elements. In: Welch, B.J. (ed.) Light Metals 1998, pp. 257–264. TMS, Warrendale, PA (1998)
121. Dionne, M., Esperance, G.L., Mirtchi, A.: Wetting of TiB2-carbon material composite. In: Eckert, C.E. (ed.) Light Metals 1999, pp. 389–394. TMS, Warrendale, PA (1999)
122. Xue, J.L., Oye, H.A.: Sodium and bath penetration into TiB2-carbon cathodes during laboratory aluminium electrolysis. In: Cutshall, E.R. (ed.) Light Metals 1992, pp. 773–778. TMS, Warrendale, PA (1992)
123. Xue, J.L., Oye, H.A.: Wetting of Graphite and Carbon/TiB2 Composites by Liquid Aluminum. In: Das, S.K. (ed.) Light Metals 1993, pp. 631–637. TMS, Warrendale, PA (1993)
124. Geng, C.: Development and application of TiB2-C composite layer cathode block for aluminum reduction cell. In: Light Metals 2001, vol. 2, pp. 50–52 (2001)
125. Ren, B.J., Xu, J.L., Shi, Z.N., et al.: Application of TiB2 coating cathode blocks made by vibration molding for 300 kA aluminum reduction cells. In: Sorlie, M. (ed.) Light Metals 2007, pp. 1047–1050. TMS, Warrendale, PA (2007)
126. Cook, A.V., Buchta, W.M.: Use of TiB2 cathode material:demonstrated energy conservation in VSS cells. In: Welch, J.B. (ed.) Light Metals 1985, pp. 545–566. TMS, Warrendale, PA (1985)
127. Sekhgar, J.A., de Nora, V., Liu, J., et al.: TiB2/colloidal alumina cathode coating in Hall-Heroult and drained cells. In: Welch, J.B. (ed.) Light Metals 1998, pp. 605–615. TMS, Warrendale, PA (1998)

128. Oye, H.A., de Nora, V., Duruz, J.J., et al.: Properties of a colloidal alumina-bonded TiB2 coating on cathode carbon materials. In: Huglen, R. (ed.) Light Metals 1997, pp. 279–286. TMS, Warrendale, PA (1997)
129. Brown, G.D., Hardie, G.J., Taylor, M.P.: TiB2 coated aluminum reduction cells: status and future direction of coated cells in Comalco. In: Aluminium Smelting Conference. Queenstown, New Zealand, Nov. 26, pp. 529–538 (1998)
130. Li, Q.Y., Lai, Y.Q., Liu, Y.G., et al.: Laboratory test and industrial application of an ambient temperature cured TiB2 cathode coating for aluminum electrolysis cells. In: Tabereaux, A.T. (ed.) Light Metals 2004, pp. 327–331. TMS, Warrendale, PA (2004)
131. Beck, T., Brooks, R.J.: Electrolytic reduction of aluminum. US Patent, 5,006,209, 1991-04-09
132. Georges, B., de Nora, V.: Aluminum production cell and cathode. US Patent, 6,358,393, 2002-03-19
133. De Nora, V.: Cell for aluminum electrowinning. US Patent, 6,093,304, 2000-07-25
134. de Nora, V., Nassau, B., Jainagesh, A.S., et al.: Method for production of aluminum using protected carbon-containing components. US Patent, 5,651,874, 1997-07-29
135. Welch, B.J.: Aluminum production paths in the new millennium. JOM. **51**(5), 24–28 (1999)
136. De Nora, V., Duruz, J.J.: Cell for electrolysis of alumina at low temperatures. US Patent, 5,725,744, 1998-03-10
137. Beck, T.R., Rousar, I., Thonstad, J.: Energy efficiency considerations on monopolar versus bipolar fused salt electrolysis cells. In: Das, S.K. (ed.) Light Metals 1993, pp. 485–491. TMS, Warrendale, PA (1993)
138. Hanneman, R.E., Hayden, H.W., Goodnow, W., et al.: Report of the American Society of Mechanical Engineers, Technical Working Group on Inert Anode Technologies, pp. 1–45. The U. S. Department of Energy (1999)
139. Welch, B.J.: Inert anodes—The status of the materials science, the opportunities they present and the challenges that need resolving before commercial implementation. In: Bearne, G. (ed.) Light Metals 2009, pp. 971–978. TMS, Warrendale, PA (2009)
140. Zhang, Z., et al.: Effect of nanopowder content on properties of $NiFe_2O_4$ matrix inert anode for aluminium electrolysis. In: Suarez, C.E. (ed.) Light Metals 2012, pp. 1381–1384. TMS, Warrendale, PA (2012)
141. Xianxi, W.: Production of NiFe2O4 nanocermet for aluminium inert anode. In: Williams, E. (ed.) Light Metals 2017, pp. 1357–1364. TMS (The Minerals, Metals & Materials Society), Warrendale, PA (2017)
142. Rhamdhani, M.A., Hayes, P.C., Jak, E.: Subsolidus phase equilibria of the Fe-Ni-O system[J]. Metall. Mater. Trans. B. **39**(5), 690–701 (2008)
143. Liu, B.G., Zhou, K.C., Li, Z.Y., Zhang, D., Zhang, L.: Microstructure and DC electrical conductivity of spinel nickel ferrite sintered in air and nitrogen atmospheres. Mater. Res. Bull. **45**(11), 1668–1671 (2010)
144. Zou, Z., et al.: Effect of sintering atmosphere on phase composition and mechanical property of $5Cu/(10NiONiFe_2O_4)$ cermet inert anodes for aluminium electrolysis. In: Lindsay, S.J. (ed.) Light Metals 2011, pp. 1135–1139. TMS, Warrendale, PA (2011)
145. Wang, G., Sun, X.: Electrochemical behaviour of cermet anodes in Na_3AlF_6-K_3AlF_6based low-melting electrolytes for aluminium electrolysis. In: Sadler, B. (ed.) Light Metals 2013, pp. 1295–1298. TMS, Warrendale, PA (2013)
146. Liu, J.-Y., et al.: Phase evolution of 17(Cu-Ni)-($NiONiFe_2O_4$) cermet inert anode during aluminium electrolysis. Trans. Nonferrous Met. Soc. Chin. **21**(3), 566–572 (2011)
147. Zhong-Liang, T., et al.: Electrical conductivities of NiFe2O4 based cermets. Mater. Sci. Eng. Powder Metall. China. **10**(2), 110–115 (2005)
148. Lai, Y.Q., et al.: Effect of metallic phase species on the corrosion resistance of 17 M-(10NiO-NiFe2O4) cermet inert anode of aluminum electrolysis. J. Central South Univ. Technol. China. **13**(3), 214–218 (2006)

149. Shan-Feng, W., et al.: Effect of initial state of metallic powder on sintering and electrical conductivity of 10NiO-NiFe2O4 ceramics. Mater. Sci. Eng. Powder Metallur. China. **16**(2), 296–302 (2011)
150. Ke-Chao, Z., et al.: Enhanced sintering and molten salt corrosion behavior of nickel ferrite based cermets. Chin. J. Nonferrous Metals. **21**(6), 1348–1358 (2011)
151. He, H.-B., Xiao, H.-N., Zhou, K.-C.: Effect of additive BaO on corrosion resistance of xCu/(10NiONiFe$_2$O$_4$) cermet inert anodes for aluminium electrolysis. Trans. Nonferrous Met. Soc. Chin. **21**, 102–108 (2001)
152. Liu, J., Zhao, M., Li, J.: Study on the inert anode for Al electrolysis based on the NiFe$_2$O$_4$spinel ceramics. In: Suarez, C.E. (ed.) Light Metals 2012, pp. 1389–1393. TMS, Warrendale, PA (2012)
153. Zhang, X., et al.: Effects of the additive ZrO$_2$ on properties of nickel ferrite cermet inert anode. Chin. J. Nonferrous Metals. **14**(6), 1002–1006 (2004)
154. Du, J., Wang, B., et al.: Study on the bubble behavior and anodicovervoltage of NiFe$_2$O$_4$ ceramic based inert anodes. In: Hyland, M. (ed.) Light Metals 2015, pp. 1193–1197 (2015)
155. Yang, J.H., Hryn, J.N., et al.: Aluminum electrolysis tests with inert anodes in KF-AlF3-based electrolytes. In: Galloway, T.J. (ed.) Light Metals 2006, pp. 421–424. TMS, Warrendale, PA (2006)
156. Yangy, J.H., Hryn, J.N., et al.: New opportunities for aluminumelectrolysis with metal anodes in a low temperature electrolyte system. In: Tabereaux, A.T. (ed.) Light Metals 2004, pp. 321–326. TMS, Warrendale, PA (2004)
157. Helle, S., Pedron, M., et al.: Structure and high-temperature oxidation behaviour of Cu-Ni-Fe alloys prepared by high-energy ball milling for application as inert anodes in aluminium electrolysis. Corrosion Sci. **52**(10), 3348–3355 (2010)
158. Helle, S., Brodu, B., et al.: Influence of the iron content in Cu-Ni based inert anodes on their corrosion resistance for aluminium electrolysis. Corrosion Science, China. **53**(10), 3248–3253 (2011)
159. Nguyen, T., de Nora, V.: De Nora oxygen evolving inert metallic anode. In: Galloway, T.J. (ed.) Light Metals 2006, pp. 385–390. TMS, Warrendale, PA (2006)
160. Simakov, D.A., et al.: Nickel and nickel alloys electrochemistry in cryolite-alumina melts. In: Sorli, M. (ed.) Light Metals 2007, pp. 489–493. TMS, Warrendale, PA (2007)
161. Chapman, V., Welch, B.J., Skyllas-Kazacos, M.: High temperature oxidation behaviour of Ni-Fe-Co anodes for aluminium electrolysis. Corros. Sci. **53**(9), 2815–2825 (2011)
162. Helle, S., et al.: Ball-milled Cu-Ni-Fe-O material as inert anodes for aluminium electrolysis in low temperature KF-AlF$_3$ electrolyte. In: Suarez, C.E. (ed.) Light Metals 2012, pp. 1377–1380. TMS, Warrendale, PA (2012)
163. Chapman, V., Welch, B.J., Skyllas-Kazacos, M.: Anodic behaviour of oxidised Ni-Fe alloys in cryolite-alumina melts. Electrochim. Acta. **56**(3), 1227–1238 (2011)
164. Zhu, Y., He, Y., Wang, D.: Fe-30Ni-5NiO alloy as inert anode for low-temperature aluminium electrolysis. JOM. **63**(5), 45–49 (2011)
165. Singleton, V., Welch, B.J., Skyllas-Kazacos, M.: Influence of cobalt additions on electrochemical behaviour of Ni-Fe-based anodes for aluminium electrowinning. In: Lindsay, S.J. (ed.) Light Metals 2011, pp. 1123–1128. TMS, Warrendale, PA (2011)
166. Helle, S., et al.: Structure and high-temperature oxidation behaviour of Cu-NiFe alloys prepared by high energy ball milling for application as inert anodes in aluminium electrolysis. Corros. Sci. **52**(10), 3348–3355 (2010)
167. Goupil, G., et al.: Cold spray deposition of mechanically alloyed Cu-Ni-Fe material for the application as inert anodes for aluminium production. In: Sadler, B. (ed.) Light Metals 2013, pp. 1283–1287. TMS, Warrendale, PA (2013)
168. Ouvarov-Bancalero, V., Guay, D., Roué, L.: Mechanically alloyed Cu-Ni-Fe-Y material as inert anode for Al production. In: Sadler, B. (ed.) Light Metals 2013, pp. 1277–1281. TMS, Warrendale, PA (2013)
169. Nguyen, T.T.: Metallic oxygen evolving anode operating at high current density for aluminium reduction cells. EP patent 2,324,142 (Appl. date 1 Sept. 2009)

170. Alcorn, T.R., Tabereaux, A.T., et al.: Operational results of pilot cell test with cermet "inert" anodes. In: Christian, M.B. (ed.) Light Metals 1993, pp. 433–445. TMS, Warrendale, PA (1993)
171. Gang, Z.: Densification and strengthening of NiFe2O4-10NiO based cermet inert anode for aluminum electrolysis, pp. 48–57. Central South University, Changsha (2007)
172. Jie, L., Zhi-Gang, W., et al.: Simulation of heat balance in 5 kA grade aluminum reduction cell with inert anodes. Chin. J. Nonferrous Metals. **19**(2), 339 (2009)
173. Wang, B., Liang, F., et al.: Pilot test of aluminum electrolysis by the NiFe2O4-M inert anodes. In: Williams, E. (ed.) Light Metals 2016, pp. 429–432. TMS (The Minerals, Metals & Materials Society), Warrendale, PA (2016)
174. Xianan, L., Shengzhong, B., Yi, S.: Review on current major carbon—Free aluminum reduction technology. In: Light Metals, vol. 3, pp. 1–4 (2019)
175. Simakov, D.A., Frolov, A.V., Gusev, A.O.: Development of inert an-odes for electrolysis. In: The second international congress "Non-Ferrous Metal—2010", September 2–4, Krasnoyarsk, Russia Part VI Aluminum Reduction Technology, pp. 354–360 (2010)
176. Simoylov, D.A., et al.: Progress in the development of reduction technology with inert anodes. In: Pashkov, G.L., Polyakov, P.V. (eds.) Proceedings of Non-Ferrous Metals, Krasnoyarsk 5–7 September, pp. 363–369 (2012)
177. UC Rusal starts pilot tests of inert anodes. http://www.rusal.ru. Accessed 4 Feb 2013
178. Mann, V., et al.: Environmental aspects of UC RUSAL's aluminum smelters sustainable development. In: Chesonis, C. (ed.) Light Metals 2019, pp. 553–563. TMS (The Minerals, Metals & Materials Society), Warrendale, PA (2019)
179. Kovacs, V., Kiss, L., et al.: Comparative analysis of the environmental impacts of aluminum smelting technologies. In: Hyland, M. (ed.) Light Metals 2015, pp. 529–534. TMS (The Minerals, Metals & Materials Society), Warrendale, PA (2015)
180. Solheim, A.: Inert anodes—the blind alley to environmental friendliness. In: Martin, O. (ed.) Light Metals 2018, pp. 1253–1260. TMS (The Minerals, Metals & Materials Society), Warrendale, PA (2018)

Part III
Study on Nano-ceramic Inert Anode

Chapter 4
Nanomaterials and Nano-cermet

4.1 Introduction

In this chapter, we will briefly review aluminum electrolysis inertial anodes, especially cermet inertial anodes, and discuss the problems that hinder their wide application. Then, we put forward some suggestions to solve these problems.

Since Hall of the United States and Heroult of France proposed a patent for starting electrolytic aluminum smelting in 1886, carbon anodes have been used in modern aluminum electrolysis production. Although considerable progress and development have been made in the past 100 years, the current status of aluminum electrolysis production is still high energy consumption, carbon anode consumption, regular anode replacement, and high-cost production instability. In addition, because aluminum electrolysis is carried out in the molten alumina containing fluorine, aluminum electrolysis reacts to produce a large amount of greenhouse gas CO_2 and carcinogenic carbon fluorides according to the following reaction.

$$Al_2O_3 + 2/3C = Al + 3/2CO_2.$$

At present, each ton of metal aluminum consumes about 14,500 kW h of electricity and carbon anode of 400–500 kg and releases 1.2 t of CO_2 and a lot of harmful gases, such as polycyclic aromatic hydrocarbons, carbonyl sulfide, hydrogen fluoride, etc. This seriously pollutes the environment and deteriorates working conditions. The electrolytic aluminum industry all over the world is looking for technological innovation. This is not only energy saving and environmental protection but also higher economic benefits. The inert anode technology is considered to be able to meet this requirement.

If an inert anode is used, aluminum electrolysis will not consume the carbon anode. Therefore, it is unnecessary to produce carbon anode. The investment in building the plant will be greatly reduced, and the production will be stable because of infrequently changing anode and adjusting pole distance. This can reduce the cost and energy consumption of aluminum electrolysis. According to the first

W. Xianxi, *Inert Anodes for Aluminum Electrolysis*, The Minerals, Metals & Materials Series, https://doi.org/10.1007/978-3-030-28913-3_4

Investment Bank of Boston, Switzerland, if the inert anode can be successfully used in the electrolytic aluminum industry, the inert anode is matched with the wet cathode. This can save energy by 10–20%, reduce the production cost by 10–15%, and reduce the investment of building factory by 20–25% [1–3]. In recent years, China's electrolytic aluminum production has developed rapidly. Based on China's primary aluminum output of 32.49 million t in 2016 [4], if the inert anode project is successfully implemented, the saved annual value of electricity and carbon materials is about 10 billion US dollars. In addition, due to the use of electrolytic inert anode and no carbon anode consumption, electrolytic reaction becomes $Al_2O_3 = 2Al + 3/2O_2$. The electrolytic production releases oxygen rather than CO_2 and carcinogenic carbon fluoride; the environment is greatly improved.

Since the use of inert anodes will cause the above revolutionary changes in aluminum electrolysis industry, various countries have been engaged in the research. In recent years, aluminum electrolytic inert anodes have always been the frontier and hot spot of aluminum metallurgy research in the world [5, 6].

Reviewing the research on inert anode in aluminum electrolysis, the researchers all over the world have carried out research on metal oxide ceramics, alloys, and cermet inert anodes. The most promising application in aluminum electrolysis industry is cermet inert anode.

As mentioned earlier, Alcoa Corporation was funded by the US Department of Energy (DOE) from 1980 to 1985 for the purpose of developing and evaluating different inert anode materials. The $NiFe_2O_4$ cermet inert anode was studied systematically, and the research report and academic paper were published in 1986. The Alcoa report confirmed that the raw material composition was 17% Cu + 42.91% NiO + 40.09% Fe_2O_3. The $NiFe_2O_4$-based cermet (so-called 5324-17Cu cermet) had the best performance, with a conductivity of 90 s/cm, and the electrode shape was almost unchanged after 30 h electrolysis. It showed good corrosion resistance and conductivity in small-scale test [7, 8]. Since then, $NiFe_2O_4$-based cermet had become the most important inert anode material for aluminum electrolysis and has been widely studied.

The Pacific Northwest National Laboratory (PNNL), funded by the US Department of Energy, had continued its research on $NiFe_2O_4$-NiO-Cu cermet inert anodes since 1985, based on the results of Alcoa research. In addition, Eltech Research had developed a CEROX coating for cermet inert anodes, funded by the US Department of Energy. Since 1991, PNNL and Eltech had carried out 6 kA electrolysis test of cermet inert anode in cooperation with Reynolds Company, respectively [9]. After 25 days of electrolytic test, the main problems were the poor thermal shock resistance of the large anode, the cracking of the electrode, the serious damage of the conducting rod, etc.

Despite the failure of the above experiments, the study of $NiFe_2O_4$-NiO-Cu cermet inert anode still had been carried on. Alcoa published the short-term technical reports in 1997, 1999, and 2001. This showed that the above problems have been solved to some extent [10–12]. Ray reported in a patent in 2001 [13] that the impurity content of primary aluminum obtained by anodic electrolysis with $NiFe_2O_4$ + NiO + Cu was 0.2% Fe, 0.1% Cu, and 0.034% NiO, respectively. In

2001, Blinov [14] was electrolyzed at low temperature of 800 °C with anodic composition of 65% $NiFe_2O_4$-18% NiO-17% Cu at alumina saturation. The annual corrosion rate of inert anode was 1.4 cm. Lorentsen et al. studied the migration mechanism of impurities brought in by the inert anode material between the anode and the cathode [15]. In 1997, Blinov et al. carried out low-temperature aluminum electrolysis experiments on the inert anode [16]. The anode component they used was supplied by Alcoa, that is, the electrolysis temperature of $NiFe_2O_4$ + 18% NiO + 17% Cu was 800 °C and the anode current density was 0.2 A/cm^2. After 130 h of electrolysis test, the corrosion rate of anode was lower than 10^{-3} $g/(cm^2\ h)$. The corrosion rate of the same anode at 950 °C was greater than 8×10^{-3} $g/(cm^2\ h)$. This study showed that low-temperature electrolysis had positive significance in reducing corrosion rate of inert anodes.

In order to improve the electrical conductivity of cermet, Alcoa added Ag to $NiFe_2O_4$ + NiO + Cu cermet studied earlier, about 50–90% of nickel and iron oxides in the cermet. The content of copper and silver or copper-silver alloy should reach 30%; the copper-silver alloy contained 90% copper and 10% silver. The results showed that decreasing the temperature can improve the corrosion resistance of the electrode. The molar ratio was 8–1.0, including CaF_2 of 25% and MgF_2 of 0.25% electrolyte; the optimum electrolysis temperature was 920 °C [17].

In September 2001, Alcoa conducted a small-scale industrialization test at a smelter in Italy. In addition, it was hoped that an industrial-scale electrolytic cell would be built in the United States. Alcoa plans to increase its production capacity of inert anodes to the level that can produce the inert anodes required for one electrolytic cell per day. The first inert anode electrolyzer production series was established in 2002. According to the report to the US Department of Energy, Alcoa planned to replace inert anodes in the carbon anode electrolyzer within 2–3 years [18]. However, Alcoa later announced that they had delayed the next stage of research deployment of the inert anode. The reason is that the thermal brittleness of the material and the connection problem with the guide rod were not well solved [19].

At the beginning of the twenty-first century, China's Northeastern University, Central South University, Tsinghua University, China Aluminum Corporation, Shandong University of Technology, and Guizhou University carried out this research work [20–22]. They also encountered the problem of anode breakage, large-scale poor connection between anode and metal conductive rod, etc.

According to the results and evaluation of the above report, the cermet anode of Alcoa was the best inert anode at present. It was the most likely inert anode for industrial applications. The main problems exposed by the test are the poor thermal shock resistance of large anodes, the cracking of the electrode, the serious damage of the conducting rod, and the failure to solve the problem of the connection between the cermet anode and the guide rod. These problems were actually mechanical properties and machinability of $NiFe_2O_4$-based cermet inert anodes [23].

Cermet is a composite composed of metal or alloy and ceramics. At present, the inert anode of cermet generally takes oxide ceramics as continuous phase to form corrosion resistance and oxidation resistance.

The network and metal phase dispersion can improve the mechanical properties and conductive properties of materials. However, cermet has the main ceramic properties, such as high brittleness, easy fracture, unreliability, rough surface, being difficult to be machined, large-scale poor connection of wire, and so on. This will affect the electrical conductivity, etc. This hinders its popularization and application in aluminum electrolysis industry. Therefore, it is necessary to solve these problems on the basis of basic performance research. Then, it can be promoted and applied in industrial production [24].

Ceramics is one of the earliest materials used by human beings and plays an important role in human history. Up to now, ceramics is still an indispensable material in human life and production. Ceramic products are widely used in various fields of the national economy. Ceramics are indispensable not only in people's daily life but also in industrial production and scientific research. This is because ceramic materials have many excellent properties than other materials, such as high wear resistance, high corrosion resistance, high-temperature and high-pressure resistance, high hardness, non-aging, etc. However, ceramic materials have a major disadvantage of brittleness. The results show that the damage is not obvious under the action of external force. This serious weakness makes ceramic materials easy to cause disastrous consequences in practical use. Thus, its application is limited. How to overcome the brittleness of ceramic materials and improve the toughness of ceramic materials has long been a direction of scientists. According to Griffith's theory, the increase in fracture strength of brittle materials can only be achieved by increasing fracture toughness and reducing critical crack size. Therefore, researchers had done a lot of work in the synthesis process, focusing on how to reduce the size and number of cracks caused by the preparation process. In addition, they designed ceramics with higher toughness and reliability.

Traditionally, there were three ways to improve the fracture toughness of ceramics. The first one was by adding energy absorption units, such as wafers, whiskers, and particles to the microstructure of ceramics. Crack deflection and bridge element were used to prevent further crack propagation. The second one is by adding human wire to ceramic substrate to absorb energy through plastic deformation. The third one is that the second phase can be transformed by adding a person. The energy of crack propagation is absorbed by phase transformation. These methods play an important role in improving the toughness of ceramics, but it is very difficult to solve the brittleness problem of ceramics.

In addition to the disadvantage of brittleness, ceramic materials have other disadvantages. For example, ceramic materials are extremely difficult to process, because ceramics are brittle materials. The hardness is higher than other materials. It is difficult to cut, plane, drill, and so on. The sintering temperature of ceramics is very high, and the equipment is expensive. The energy consumption is large, which is not conducive to the protection of the environment. How to make ceramic materials sintered at low temperature with good processability is a problem that many materials researchers have tried to solve for a long time, and have not been solved yet. However, the emergence of nano-ceramics has brought new hope for solving all the abovementioned problems. It overcomes the brittleness of ceramic materials, processing difficulties, sintering temperature, high energy consumption, and other disadvantages.

4.2 Nanomaterials

Nanometer (nm) is a unit of measure, 1 nm = 10^{-9} m. Nanomaterials are materials with at least one phase grain size less than 100 nm. Until the concept of mesoscopic was put forward in the 1980s, man's understanding of the material world entered a new stage, that is, the mesoscopic field between macro and micro. It includes clusters, submicron, and nanometers in the middle. The range of nanosystems is from 1 to 100 nm.

Nanotechnology is a new technology that was born and developed in the late 1980s. Its basic meaning is to recognize and transform nature in the nanometer size range. It involves the non-macro and non-micro intermediate field, which was rarely involved in the past, thus opening up a new level of human understanding of the world. This marks that human science and technology has entered a new era—the era of nanotechnology. It will add an important new connotation to the scientific and technological revolution. Nanoparticles usually contain only a limited number of units and no longer have periodic conditions. Surface atoms account for a large proportion of the surface structure and are different from the internal thermal motion. The electron motion and electron energy level of nanoparticles are different from bulk solids. Nanocrystalline particles and their composed solids have small size effects, surface and interface effects, quantum size effects, and special properties that many traditional solids derived therefrom do not possess, including high strength and high hardness. Good plasticity and toughness are attractive properties of nanomaterials.

4.3 Nano-ceramics

Nano-ceramics are important nanomaterials, which refer to ceramic materials consisting of single-phase nanocrystalline particles or ceramic materials with nanoscale microstructures. Its grain size, grain boundary width, pore size, defect size, and so on were on the order of magnitude. Nanocrystalline ceramics have become one of the three trends in ceramic research. The appearance of nano-ceramics can overcome the disadvantages of brittleness, processing difficulties, high sintering temperature, high energy consumption, and so on. This brings new hope to solve all the above ceramic problems.

In 1987, Karch et al. from Germany first reported for the high toughness and low-temperature superplastic behavior of the nano-ceramics. Figure 4.1 shows the plastic deformation of the nano CaF_2. First, the flat square sample was placed between two pieces of aluminum foil, and the pressure was applied along the arrow

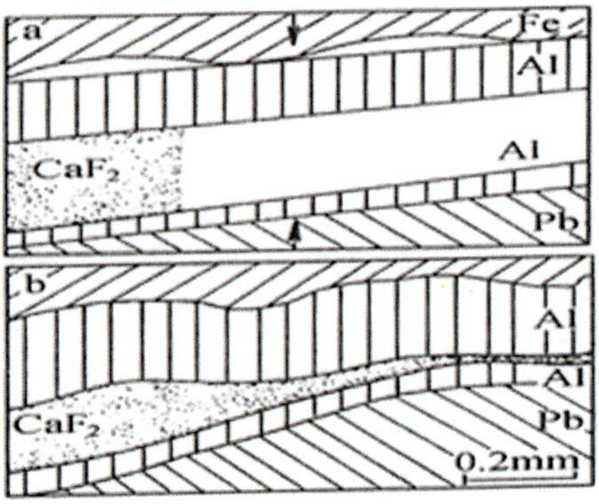

Fig. 4.1 The plastic deformation of nanometer CaF_2 [25]

direction to close it up and down. It was found that the shape of the sample was sinusoidal bent due to plastic deformation of the nanocrystalline CaF_2. It became filaments by plastic flow to the right. In addition, they placed the non-compact nano-ceramics produced by vacuum pressing titanium oxide nanoparticles with an average diameter of about 8 nm into a special mold and loaded it at 180 °C for 1 s. As a result, the plate specimen was bent 180° without crack propagation. These results showed the potential excellent properties of nano-ceramics for the first time and provided a new way to solve the brittleness problem of ceramics. This had been puzzling people for a long time. "Nano-ceramics is a strategic way to solve the brittleness of ceramics," Kahn, a famous British materials scientist, published in *Nature*.

Since then, countries all over the world have high hopes on the development of nano-ceramics to solve the brittleness and difficult-to-process properties of ceramic materials. Scientists have done a lot of research and achieved fruitful results in structure, properties, etc. For example, many studies had proved that the sintering properties of nano-ceramic powders were very good, and their sintering temperature was several degrees lower than that of ordinary ceramic powders. The nanocrystalline TiO_2 ceramics had extremely high hardness at 100 °C. The microhardness of nano-sized TiO_2 ceramics was 1300 kgf/mm^2 (1 kgf = 9.8 N), and the microhardness of ordinary TiO_2 ceramics was less than 200 kgf/mm^2. In the aspect of characterization of nanocrystalline ceramics, the grain boundary structure of nanocrystalline ceramics was studied as the starting point, advanced techniques, such as high-resolution electron microscope, electron diffraction, neutron scattering, X-ray photoelectron spectroscopy, and scanning tunneling microscope, and had been applied more widely.

In addition, many studies have shown that nano-ceramics not only have the excellent properties of advanced ceramic materials, such as high wear resistance, high corrosion resistance, high-temperature resistance, high hardness, and being not easy to aging, but also had unique properties, such as high toughness and superstability. Nano-ceramics may be forged, extruded, pulled, and bent like any other materials and can be used to fabricate parts of precision size directly. High toughness changes the fragile image of ceramics [25].

According to B. R. Lawn [26], the concept of brittleness index (*B*) was proposed. A. R. Boccaccin [27] used brittleness index to evaluate the machinability of materials, brittleness index = hardness/fracture toughness. The brittle indexes of steel and monocrystalline silicon are 0.1 $\mu m^{-1/2}$ and 17 $\mu m^{-1/2}$, respectively. The brittleness index of glass and ceramics is generally between 3 and 7 $\mu m^{-1/2}$. According to the research situation of nano-ceramics in many countries, the ductility, especially fracture toughness of nano-ceramics, decreases its brittleness index. The machinability of nano-ceramics can be improved, which is close to the machinability of iron and steel.

As an important part of nanomaterials, nano-ceramics undoubtedly plays a unique role in the industrial revolution in the new century.

To sum up, the properties of nano-ceramics are most likely to be applied in the following aspects.

Room-temperature superplasticity is one of the most attractive potential properties of nano-ceramics and one of the most promising aspects of nano-ceramics. Common ceramic materials are too hard and too brittle to be machined. Therefore, it is very difficult to cut and drill holes like metals. This limits the application of common ceramic materials.

The room-temperature superplasticity of nanocrystalline ceramics makes it possible for ceramics to be calcined, extruded, drawn, and bent as other materials, without grinding, on the premise of retaining their excellent properties, such as high chemical corrosion resistance, high-temperature and high-pressure resistance, etc. They can be directly prepared as precision size parts.

High toughness is another excellent potential property of nano-ceramic. Ceramic materials have excellent wear resistance and corrosion resistance. Because of their inherent brittleness, they always exist in people's minds as "fragile products." They have narrower application field than mainstream materials, such as steel and plastics. The appearance of nanocrystalline ceramics may completely change this image.

The traditional ceramic industry has the characteristics of high energy consumption and serious pollution. This is obviously a drawback in an era of increasingly tight energy and serious environmental damage. However, the present research shows that nano-ceramic can be sintered at a temperature below 1000 °C, and many nano-ceramic have been densified below 1000 °C, which may continue to be significantly reduced. This not only saves a lot of valuable energy but also protects the environment. Therefore, the use of nanotechnology to transform the traditional ceramic industry may lead to another revolutionary development of the ceramic industry. The application of low-temperature sintering properties of nano-ceramics is not limited. For example, it has been found that the properties of sintering at low temperature can be used to bond ceramic materials, and the ordinary Y-TZP ceramics are joined together by using nano-sized $ZriO_2$ (3Y) powders. Compared with the common brazing or partial liquid phase bonding, nano-ceramic bonding can ensure that the bonding layer is the same as the ceramic substrate. Therefore, the physical and chemical properties and even the mechanical properties of the ceramic materials are consistent. This kind of nano-ceramic based on low-temperature sintering properties will be applied more and more widely.

In addition to the structural and performance-based applications listed above, another important application of nano-ceramic is functionality. The electrical, magnetic, optical, and acoustic properties of ceramic materials are usually closely related to their grain boundary phases. Because of the small grain size, the proportion of grain boundary phase in unit volume is much higher than that in common coarse grain ceramics, and the composition of grain boundary phase may be very different from that of ordinary coarse grain ceramics. Therefore, it is possible to bring many unexpected changes to the properties of ceramics. The nonlinear voltammetric characteristics of nano-ZnO ceramics have verified this view [28].

The application of nano-ceramic is wider than that. Through the deep research, more excellent properties will be found and widely used. These unique properties of nano-ceramic make it possible to solve the problem of cermet inert anode application in aluminum electrolysis.

Compared with other nanomaterials that used commercially, although nano-ceramic have been studied for nearly 30 years. However, because the system was not yet complete in theory, it lacks the experience accumulation. So far, no breakthrough has been made in commercial application, and there are many problems to be solved.

In recent years, some researchers have used nanomaterials to add materials for preparing inert anodes in nano-cermet and alloys containing nanoparticles for experiments on inert anodes [146, 147, 172 of Chap. 3]. The experimental results show that nanomaterials can be used in aluminum electrolytic inert anodes of metal alloys or cermet and improve the compressibility, density, bending strength, and impact toughness of nano-cermet materials. This helps to solve the problem of cermet inert anodes.

4.4 Nanocomposite Ceramics

4.4.1 Definitions of Nanocomposite Ceramics

Niihare et al. found that the mechanical properties at room temperature and high temperature can be greatly improved by adding superfine particles into Al_2O_3 and MgO. They call this material nanocomposite or nanocomposite ceramics [29].

Many experiments show that, compared with the case of traditional materials, the mechanical properties of nanostructured materials are significantly changed and the strength and hardness of some materials are increased exponentially. For oxide ceramics such as Al_2O_3, MgO, and mullite, the strength and toughness are low, and the mechanical properties at high-temperature and thermal shock resistance are poor. This is expected to be overcome by adding the second phase (such as particles, whiskers, fibers, lamellar crystal, etc.).

In order to solve the brittle fracture of ceramics, the toughening of ZrO_2 whisker particles based on nanocomposite not only exerts the advantages of both high strength of nanoparticle and high toughness of micron composite but also have synergistic effect to obtain nano-micron composite materials. For example, adding 2.5Y-ZrO_2 into the Al_2O_3/SiC nanosystem can increase the fracture toughness and strength by 30% and 70%, respectively. In the practical application of nano-ceramics technology, they in Japan has been used for cutting tools, sealing rings, and in United States is ready for wear and corrosion resistant components.

At the end of the 1990s, the Japanese New Harahara took the lead in the research of Si_3N_4/BN nano-multiphase machinable ceramics. The strength of this kind of machinable ceramics was above 1000 MPa, achieving the high strength and machinability. Since then, some researchers have further systematically studied nanocomposite machinable ceramic systems such as Si_3Nt/BN, SiC/BN, Al_2O_3/BN, and AlN/BN, and made remarkable progress. Through nanocomposites, ceramics not only have high strength and toughness but also have good machinability. This can overcome the disadvantages of low strength, high brittleness, and being difficult to be machined. It provides a new way for the design of high performance and machinable ceramics.

Some progress has been made in the study of machinable ceramics in China. For example, Zhang Jian of Wuhan University of Technology has purified and removed impurities and controlled the fineness of common ceramic materials. After proportioning and homogenization, zirconia nanoscale dispersion strengthening and toughening technology was adopted in the process of ceramic preparation, so that the flexural strength of the material was increased from 150 to 230 MPa. The precision machined ceramics can be machined by conventional machining methods, such as turning, milling, drilling, planing, and grinding. The precision machining accuracy of the ceramics can reach 0.01 mm [30].

It has been reported that the room-temperature strength and toughness of nanocomposite ceramic materials are 2–5 times higher than that of single-component materials. The hardness, toughness, strength, creep resistance, fatigue resistance, and thermal shock resistance of nanocomposite ceramics are greatly improved at high temperature. As mentioned above, by adding ultrafine particles to Al_2O_3, MgO oxide ceramics called nanocomposites, the mechanical properties of Al_2O_3, MgO oxide ceramics can be greatly improved at room temperature and high temperature. This breakthrough makes people find the new strategic way of ceramic toughening.

4.4.2 Classification of Nanocomposite Ceramics

The nanocomposite ceramic materials can be classified into four categories, i.e., intragranular, intergranular, intragrain/intergranular, and nanometer/nanocomposite materials, as shown in Fig. 4.2 [31]. In intragranular and intergranular nanocomposites, nanoparticles are mainly dispersed in or between matrix grains. It improves not only the properties and durability of the materials at room temperature but also the mechanical properties of the materials at high temperature, such as hardness, strength, creep resistance, and fatigue failure resistance. Nanocomposites are

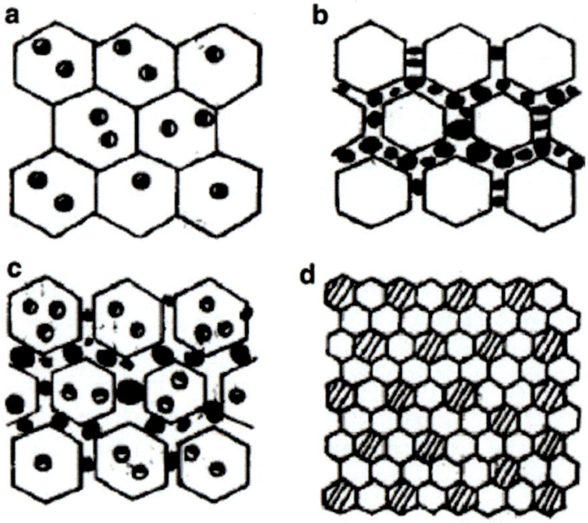

Fig. 4.2 Structural types of nano-ceramic composites. (**a**) Intragranular. (**b**) Intergranular. (**c**) Intragrain/intergranular. (**d**) Nanometer/nanocomposite materials

composed of nano-dispersion and matrix grains, which will add some new functions to ceramic materials. The machinability and plastic resistance of nanocomposites are similar to that of metals.

From the understanding of the microstructure of the material, the grain size in the material is directly related to the properties of the material. When the grain size of ceramic material reaches nanometer order, the properties of ceramic material will have obvious abnormal performance. The existence of nano-ceramic superplasticity is an example. When the grain size of tetragonal zirconia ceramics is as small as 130 nm, at 1250 °C (less than half of the melting point of zirconia), such as bending, pressing, and shearing, it can deform up to 400% under different stresses. Nano-ceramic material has obvious superplastic behavior at medium temperature, and the brittleness of ceramic material is changed.

The same ceramic material was used for tensile fatigue test at room temperature. After the fracture of the ceramic materials, the zirconia grains in the fracture layer from the surface to the following 2 μm were elongated, and the deformation degree reached 300–400, while the inner layer was not deformed. This means that at room temperature, nanocrystalline ceramic materials exhibit superplastic behavior in microdomains. These unusual behaviors of nanocrystalline ceramic materials are at least enlightening and thinking. If the grain size of ceramic materials reaches nanometer size, the superplastic behavior of ceramic materials is widespread. The problem of brittleness of ceramic materials is expected to be solved [32, 33]. However, it is not easy to make the grain size of ceramic materials to reach nanometer size. This is because the preparation of nano-ceramic materials is not easy; it tends to the research of nanocomposite ceramic materials.

The strength of Si_3Ne ceramics modified by nanometer SiC or Si_3Ne is much higher than that of conventional materials. British scientists added nano-$A1_2O_3$ to ZrO_2 to obtain high-toughness ceramics, and the sintering temperature decreased about 100 °C. Germany's Ulrich found that the impact toughness of ceramics prepared by mixing 20% nanometer SiC powder into coarse crystalline SiC increased the impact toughness by 75%. In China, nano-Al_2O_3-modified Al_2O_3 substrate was fabricated, which has nearly doubled the impact toughness, and the strength and toughness of 85 and 95 ceramics modified by nano-Al_2O_3 were increased by more than 50%.

The preparation of nano-modified materials is an effective, economical, and operational way to strengthen and toughen conventional materials (especially brittle materials such as ceramics).

4.5 Nano-cermet

4.5.1 Definition of Nano-cermet

Nano-cermet is also a kind of nanocomposite ceramics, and nanocomposite ceramics refers to the ceramic matrix containing nanoparticles of the second-phase composite materials. Nanocomposite ceramic is a kind of high-tech ceramic material, which is the extension and development of nanomaterial. Nano-cermet is a new type of ceramic, which can refine grain and optimize the mechanical properties of materials by adding nanomaterials into traditional cermet.

4.5.2 Design of Nano-cermet Inert Anode

In recent years, the inert anode of aluminum electrolysis has become the research frontier and hot spot of aluminum metallurgy research in the world. The use of inert anodes no longer consumes carbon anode, and the stability and shortening polar distance can greatly reduce the electrical consumption of electrolytic aluminum production. Thus, there is no harmful gas CF compounds and CO_2 in electrolytic aluminum production, which is beneficial to the environmental protection. It may cause the revolutionary change of the aluminum industry, so that many countries carry out the research competitively.

At present, the best performance and the most promising material for anode is the $NiFe_2O_4$ cermet developed by Alcoa Corp. since the 1980s. However, due to the brittleness of $NiFe_2O_4$ cermet anode material, some mechanical properties, such as strength, toughness, and ductility, are poor, which limits the use of it [34–41]. In recent years, the researches on nano-ceramics has shown that the introduction of nano-dispersed phase into the micron grade matrix for compounding can improve the mechanical properties of the ceramic materials, which has attracted widespread attention of the materials scientists [32, 33]. On the basis of our study on $NiFe_2O_4$ cermet [42, 43], the $NiFe_2O_4$ nano-cermet can be obtained by modifying with nano-NiO.

The $NiFe_2O_4$ nano-cermet is similar to $NiFe_2O_4$ cermet in corrosion resistance and conductivity through experimental tests. This can be used as an inert anode for aluminum electrolysis. In addition, lowering of the sintering temperature and increasing of the density, strength, and toughness are beneficial to improving the processability of the cermet.

Some experimental results about our preparation, corrosion, conductivity, and research contents about the improvement of mechanical properties and processability will be published here.

Nano-cermet is a kind of nanocomposite ceramics. Nanocomposite ceramics refers to the ceramic matrix containing nanoparticles of the second-phase composite materials. Nanocomposite ceramic is a kind of high-tech ceramic material, which is the extension and development of nanomaterial. Nano-cermet is a new type of ceramic, which can refine the grain and optimize the mechanical properties of materials by adding nanomaterials into traditional cermet.

Through the investigation of inert anode, it is found that $NiFe_2O_4$-based cermet with spinel structure is the most likely substitute for aluminum electrolytic carbon anode at present (Fig. 4.3).

Spinel composite with excellent thermal and chemical stability is the most promising ceramic matrix material for aluminum electrolytic inert anode.

Spinel oxides are ionic compounds, and the general formula is that the AB_2O_4 spinel structure system is connected by equiaxed unit cells to form a framework. This structure is reflected in the morphology, usually in the intact octahedron spinel structure; A-O and B-O are strong ionic bonds. The electrostatic strength of each bond is equal, and the structure is firm, so it has good chemical stability. It has strong resistance to various melt corrosion at high temperature. Because it belongs to cubic system, its thermal conductivity and thermal expansion are the same in all directions. The spinel oxide materials can be used as electrode materials in molten salt system because of their small expansion coefficient and good thermal stability.

Nanometer NiO is a kind of green to dark green powder, and it becomes yellow powder when it is overheated. NiO belongs to NaCl-type cubic crystal. There are six

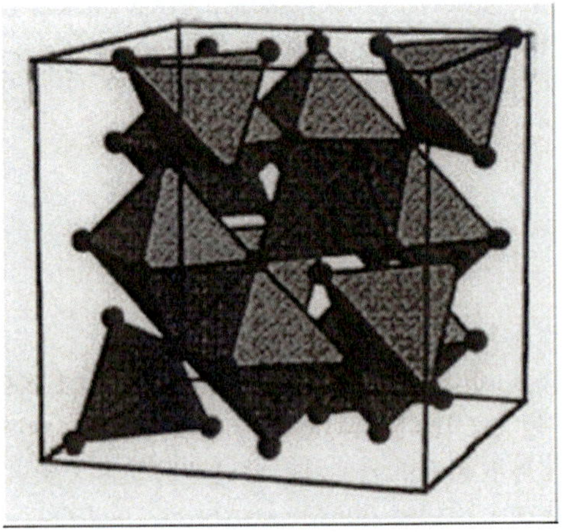

Fig. 4.3 Spinel structure

O atoms around Ni atom and six Ni atom around O atom. Their coordination numbers are all six. Because the type of polyhedron mainly depends on the ratio of positive and negative radius, the radius of Ni_{2+} is 69 pm and that of O_{2-} is 140 pm. The ratio of positive and negative ions is 0.507, which is greater than 0.414. It is concluded that NiO is an octahedral coordination. The melting point of NiO is 1984 °C, soluble in acid and ammonia water and insoluble in water.

We had studied the inert anode material of cermet made with 60% NiO + 40% Fe_2O_3 + 17% Cu and proved that it has superior electrical conductivity and good corrosion resistance at high temperature. However, it is the same as other ceramic materials, which are difficult to process in large scale. Based on previous research, we intend to add nanometer NiO and optimize the preparation process to make nano-sized NiO-Cu-$NiFe_2O_4$ cermet. The toughening method of this material is mainly toughened by nanoparticles and supplemented by phase change toughening. It is hoped that NiO-Cu-$NiFe_2O_4$ nanocomposite cermet can be used as electrolytic inert anode and solve the problems of large-scale poor machinability of inert anode.

4.6 Process for Preparing Nano-cermet Inert Anode

The selection of adding nanomaterials must take into account the similar doping of the nanomaterials. The oxide doping of inert anode cermet material is mainly focused on the selection and the best content of the oxide sintering aids. The aids are

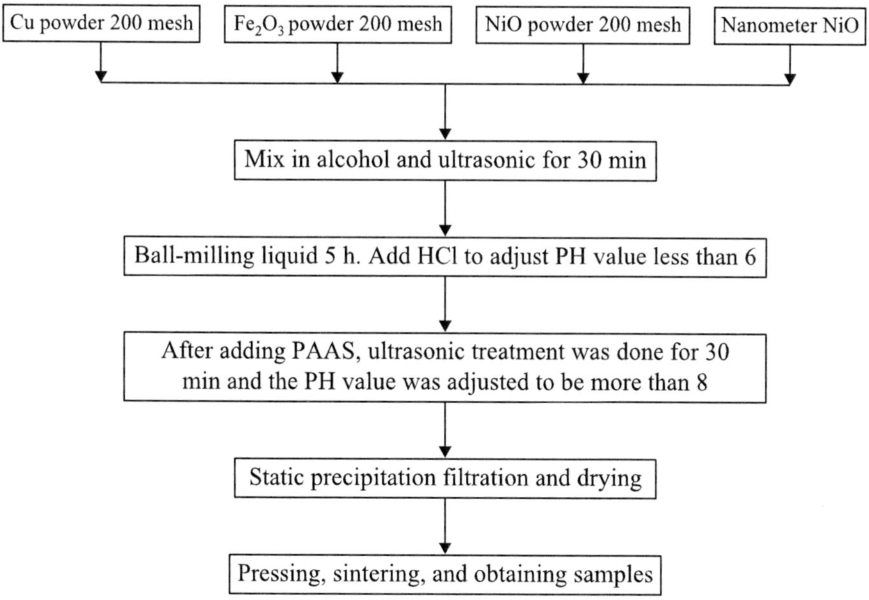

Fig. 4.4 The flow of samples preparation

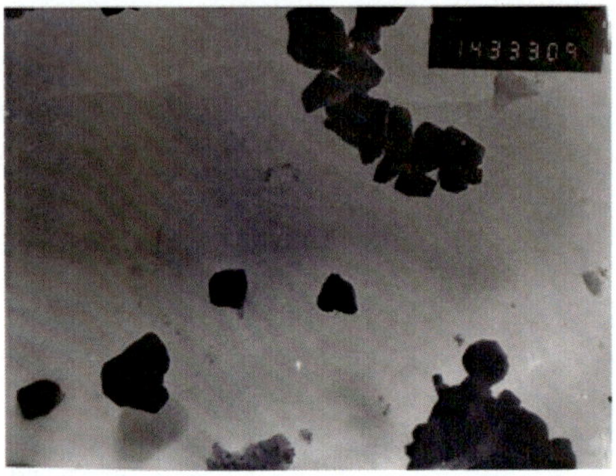

Fig. 4.5 TEM photo of nano-NiO (1 × 143,000)

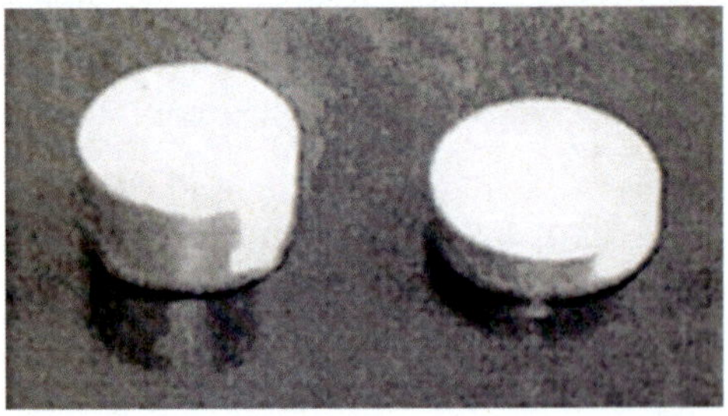

Fig. 4.6 Preparative nano-cermet (D = 3 cm)

required to increase the density of material, have lower solubility in cryolite molten salt, and not react with the molten salt [44]. In order to satisfy the above requirements, nano-NiO was selected as the additive in this paper.

The specific preparation process is as follows (percentage of nanometer NiO to total (NiO + Fe_2O_3)%) (Figs. 4.4, 4.5, and 4.6).

The microstructure of $NiFe_2O_4$ nano-cermet was analyzed by X-ray diffraction method. The density of the nano-cermet was measured by Archimedes method.

The resistance test of the nano-cermet is as follows. The sample was placed in a high-temperature energy-saving box-type resistance furnace, measured with the ZL5 type of intelligent LCR measuring instrument (bridge), and then converted to the conductivity. The nano-cermet samples were placed in the molten aluminum electrolyte and corroded under 1000–1050 °C.

After preparing the metallographic samples well, the surface microstructure of the sample was observed with OLYMPUS BX 10M metallurgical microscope.

4.7 X-Ray Diffraction Analysis

To detect whether the prepared sample is nano-cermet according to Scherrer formula, the full width at half maximum (FWHM) $\beta_{1/2}$ of the diffraction peak is used to calculate the grain size D, which is perpendicular to the crystal surface. If the diffraction peak width is only caused by the grain refinement, and the grain is uniform, the formula is as follows:

$$D(h,l,k) = \frac{k\lambda}{\beta_{1/2}\cos\theta}$$

where D is the grain size in the vertical direction along the plane (h,l,k); k is the Scherrer constant, and its value is 0.89; $\beta_{1/2}$ is FWHM of the diffraction peak; and λ is the wavelength of the copper target of 0.15406 nm.

By the characteristic of three-intensity peaks, checking PDF cards to find the corresponding FWHM and diffraction angle, the grain size was calculated according to Scherrer formula. The XRD pattern of the sample of nano-NiO 3% is shown in Fig. 4.7 (Figs. 4.8 and 4.9, Tables 4.1, 4.2, and 4.3).

It can be known from the diffraction pattern analysis that the most substance of the samples is $NiFe_2O_4$, as well as part of NiO and $Cu_{0.2}Ni_{0.8}O$. In addition, there are some CuO, indicating that the target product $NiFe_2O_4$ is produced, and the composition of the sample is relatively simple.

The mean diameter calculated of the nano-NiO grains is in the range of 20–30 nm. According to the definition of the nanocomposite ceramics, $NiFe_2O_4$ nano-cermet was obtained in the experiment.

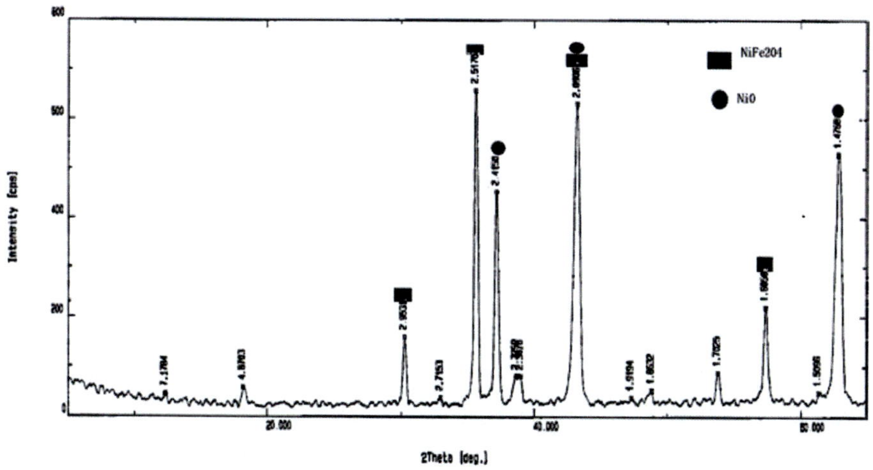

Fig. 4.7 XRD pattern of the sample of nano-NiO 3%

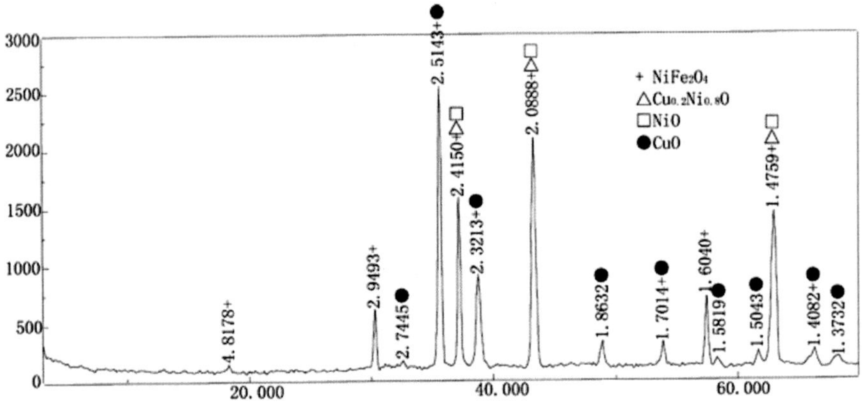

Fig. 4.8 XRD pattern of the sample of nano-NiO 5%

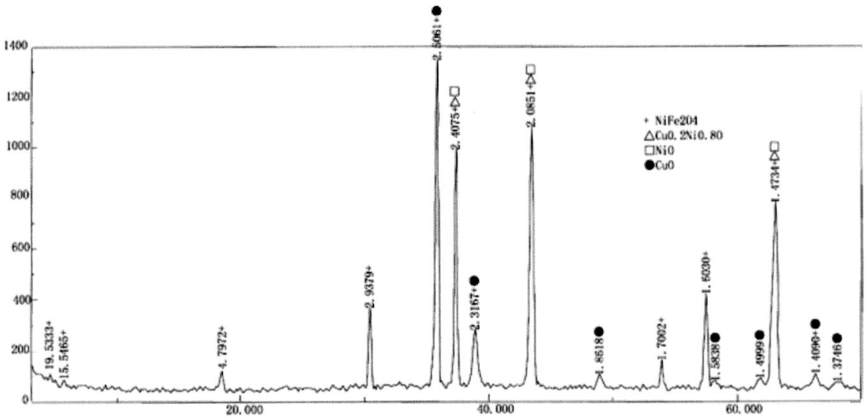

Fig. 4.9 XRD pattern of the sample of nano-NiO 7%

Table 4.1 The intensities, FWHMs, and grain sizes of nano-NiO 3%

Corresponding crystal planes of NiO	Intensities (I)	FWHMs ($\beta_{1/2}$)	Diffraction angles (θ)	Grain sizes (D/nm)
(111)	2.4150	0.00574	18.60	25.20
(200)	2.0906	0.00812	21.62	17.96
(200)	1.4768	0.00272	31.44	31.81

Table 4.2 The intensities, FWHMs, and grain sizes of nano-NiO 5%

Corresponding crystal planes of NiO	Intensities (I)	FWHMs ($\beta_{1/2}$)	Diffraction angles (θ)	Grain sizes (D/nm)
(111)	2.4075	0.00539	18.66	25.21
(200)	2.0833	0.00742	21.70	19.98
(220)	1.4751	0.00269	31.48	32.05

Table 4.3 The intensities, FWHMs, and grain sizes of nano-NiO 7%

Corresponding crystal planes of NiO	Intensities (I)	FWHMs ($\beta_{1/2}$)	Diffraction angles (θ)	Grain sizes (D/nm)
(111)	2.4225	0.00656	18.54	20.90
(200)	2.0888	0.00574	21.64	25.34
(220)	1.4751	0.00410	31.48	38.00

Nano-NiO particle dispersion will suppress the oxide matrix grain formation and reduce the abnormal grain growth, thereby making the grains refined. The grain refinement is beneficial to the slip between the grains, which can enhance the plasticity of the material. The addition of nano-NiO particles plays a role in improving the mechanical properties of the cermet.

4.8 Metallographic Structure Observation

Taking the polished samples, we labeled them as AN0 (nano-NiO addition amount was 0%), AN3 (nano-NiO addition amount was 3%), AN5 (nano-NiO addition amount was 5%), and AN7 (nano-NiO addition amount was 7%). Each sample was corroded in nitric acid alcohol solution (HNO_3 of 2 ml + 95% alcohol of 100 ml) for 1 min. After metallographic sample was prepared, the surface microstructure of the sample was observed with OLYMPUS BX 10M microscope. The metallographic structures of the samples are shown in Fig. 4.10.

The black parts of the metallographic photos should be a large number of spinel $NiFe_2O_4$ ceramic phase, which contains NiO and Cu phases. The white bright area is the metal phase, which may be produced in the sintering process. The ceramic phase is unevenly distributed in the sample AN0 (no nano-NiO), a small amount of ceramic phase is distributed in the metal phase, and the porosity of the material is *high*. When adding nano-NiO, such as sample AN3 (containing nano-NiO 3%), there are larger ceramic particles, indicating that the sintering effect of metallic phase and ceramic phase is not good. With the increase of nano content, it can be seen from the picture of sample AN5 (containing nano-NiO 5%) that the ceramic phase is uniformly distributed; the metallic and ceramic phases present network-like distribution with higher density. This indicates that both had been sintered. From the picture of the sample AN7 (including nano-NiO 7%), it can be seen that too much ceramic phase causes a decrease in the metallic phase of the ceramic phase.

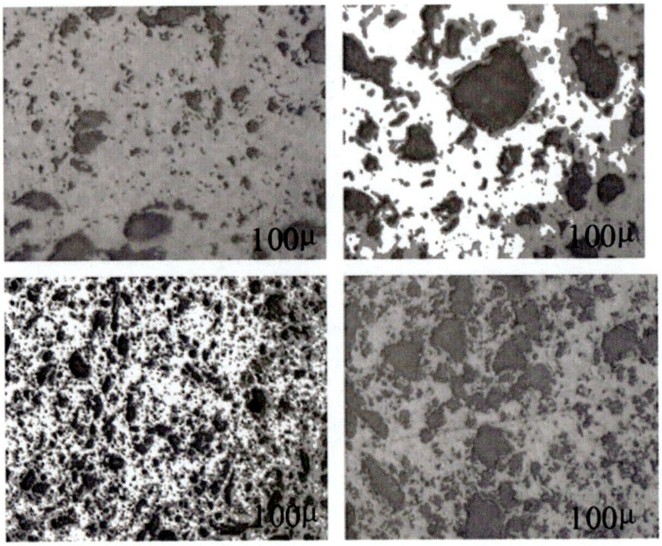

Fig. 4.10 Metallographic pictures of samples of AN0–AN7

Fig. 4.11 The device for resistivity measurement. (1) Platinum wire; (2) platinum sheet; (3) sample; (4) refractory bricks; (5) controllable temperature resistance furnace

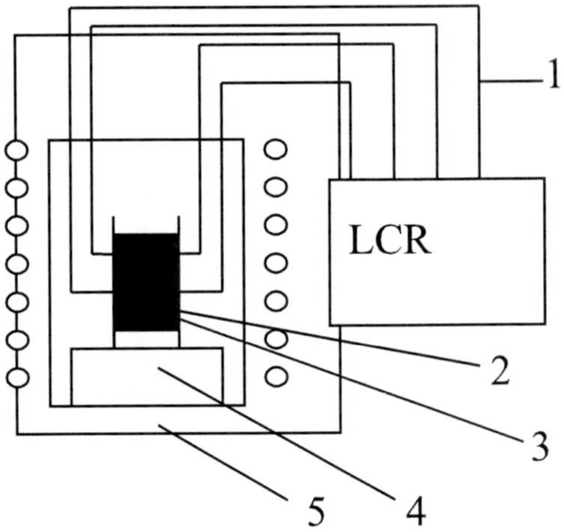

4.9 The Determination of Electrical Conductivity

AN3 used in the test experiment is a ZL5-type intelligent LCR measuring instrument (bridge), using DWT-702-type temperature control instrument with thermocouple to control furnace temperature (Fig. 4.11).

Calculation formula of electrical conductivity:

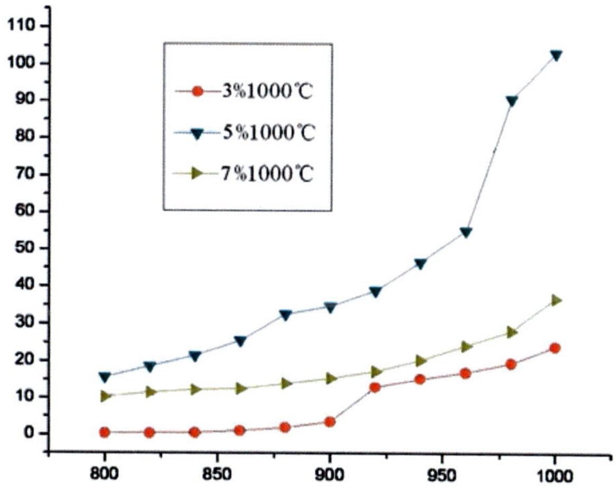

Fig. 4.12 Effect of nano-NiO content on electrical conductivity. *x*, temperature (°C); *y*, conductivity (1/Ω cm)

Table 4.4 The sample (L = 10.00 mm, D = 14.00 mm) conductivity of nano-NiO 3%

Temperatures (°C)	Resistance values (Ω)	Quality factors	Resistivity (Ω cm)	Conductivity (1/Ω cm)
800	1.7587	119	2.70616	0.3695
820	1.2259	137	1.88635	0.5301
840	1.1106	150	1.70892	0.5852
860	0.5984	152	0.92083	1.0860
880	0.3336	179	0.51334	1.9480
900	0.1863	185	0.28662	3.4889
920	0.05026	208	0.077338	12.9303
940	0.04276	205	0.065796	15.1985
960	0.03817	728	0.058732	17.0265
980	0.03332	649	0.051273	19.5034
1000	0.02713	413	0.041747	23.9538

Table 4.5 The conductivity of sample (L = 9.00 mm, D = 12.00 mm) of nano-NiO 5%

Temperatures (°C)	Resistance values (Ω)	Quality factors	Resistivity (Ω cm)	Conductivity (1/Ω cm)
00	0.04585	70	0.063992	15.6269
820	0.03850	75	0.053736	18.6095
840	0.03335	79	0.046543	21.4855
860	0.02803	50	0.039119	25.5630
880	0.02199	46	0.030694	32.5797
900	0.02062	39	0.028783	34.7427
920	0.01841	36	0.025691	38.9241
940	0.01538	40	0.021469	46.5788
960	0.01295	154	0.018079	55.3128
980	0.007902	213	0.011029	90.6701
1000	0.006935	209	0.0096793	103.3133

Table 4.6 The conductivity of sample (1000 °C, $L = 6.00$ mm, $D = 14.00$ mm) of nano-NiO 7%

Temperatures (°C)	Resistance values (Ω)	Quality factors	Resistivity (Ω cm)	Conductivity (1/Ω cm)
800	0.02298	90	0.098223	10.1809
820	0.02040	91	0.087162	11.4729
840	0.01917	81	0.081933	12.2051
860	0.01871	81	0.079964	12.5056
880	0.01677	67	0.071665	13.9538
900	0.01522	59	0.065057	15.3711
920	0.01354	61	0.057872	17.2795
940	0.01154	67	0.049328	20.2725
960	0.009652	63	0.041247	24.2442
980	0.008318	52	0.035547	28.1318
1000	0.006355	91	0.027157	36.8229

$$\sigma = \frac{1}{p} = \frac{L}{SR} = \frac{L}{\pi r^2 R}$$

where σ is the conductivity (1/Ω cm), ρ is the resistivity (Ω/cm), R is the resistance (Ω), L is the thickness (cm), and S is the cross-sectional area (cm^2) (Fig. 4.12, Tables 4.4, 4.5, and 4.6).

The conductivities of the samples with different contents of nano-NiO sintering at the same temperature are not proportional to the amount of addition. As the NiO content increases from 3% to 7%, the conductivity of the sample at 800–1000 °C is the lowest first and then begins to increase. When the adding amount is 5%, electrical conductivities reach to 90.6701 s/cm at 980 °C and 103.3133 s/cm at 1000 °C. Therefore, the addition of nano-NiO can improve the conductivity of the cermet, and the optimal amount is about 5%. The electrical conductivity of nano-NiO-Cu-$NiFe_2O_4$ cermet material shows the rising trend with the increase of temperature, with the properties of high-temperature semiconductors.

In addition, the conductivity of the material has a great relationship with the density of the material. The lower the density, the more the pores in the material. The pores make the conductive phase of the material disconnected and hinder the movement of the conductive particles, thereby reducing the conductivity of the material.

From the previous metallographic analysis photos, we find that the ceramic phase of AN5 (containing nano-NiO 5%) is uniformly distributed, metallic and ceramic phases present network-like distribution, the density of the sample is higher, and the electrical conductivity is higher.

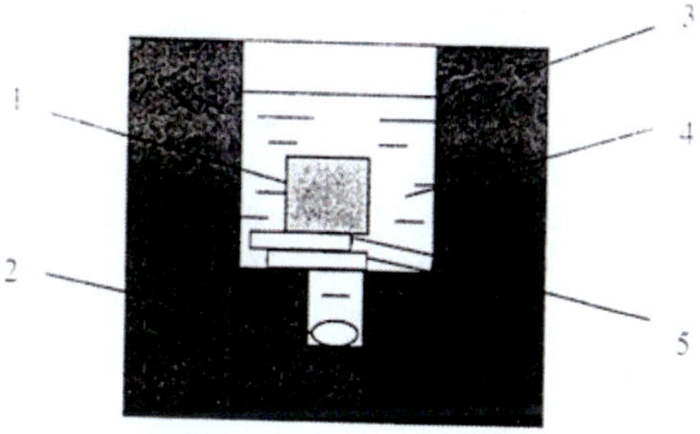

Fig. 4.13 Corrosion device of nano-cermet. (1) Nano-cermet; (2) aluminum liquid; (3) graphite crucible; (4) aluminum electrolyte; (5) alumina

Table 4.7 The corrosion test conditions

Electrolyte	Corrosion temperatures	Corrosion time
$2.7Na_2AlF_6 + 4\%AlF_3 + 5\%CaF_2 + 6\%Al_2O_3$	1000–1050 °C	12 h

Table 4.8 The corrosion test results

Samples	Corrosion time (h)	Volume before corrosion (cm^3)	Volume after corrosion (cm^3)	Area before corrosion (cm^2)	Area after corrosion (cm^2)	Annual corrosion rate (cm/year)
B	12	2.211	2.171	13.118	13.054	2.22
C	12	3.923	3.852	16.125	15.834	2.31

4.10 Corrosion Rate

The device for measuring the corrosion rate of nano-cermet is shown in Fig. 4.13. The corrosion experiments were carried out in molten aluminum electrolyte, and the experimental conditions are shown in Table 4.7. The corrosion test results are shown in Table 4.8.

The corrosion rate in Table 4.7 were calculated by the formula:

$$C = 2\Delta V \cdot m / \left(S_0 + S\right) \cdot \tau \tag{4.1}$$

where C is the corrosion rate (g/cm^3 h) and ΔV is the volume change of the sample before and after corrosion. S_0 and S are areas of a sample before and after corrosion, respectively; m is the density; τ is the annual corrosion rate (cm/year). Then, we have

$$C_1 = C \cdot 24 \cdot 365 / m \tag{4.2}$$

The density of B is 5.20 g/cm^3; The density of C is 5.32 g/cm^3.

4.11 Conclusions

1. $NiFe_2O_4$ nano-cermet can be prepared by the methods of homogeneous precipitation, molding, and normal-pressure sintering.
2. The sintering temperature of $NiFe_2O_4$ nano-cermet is 1150 °C, which is 100 °C lower than that of cermet.
3. The conductivity of $NiFe_2O_4$ nano-cermet increases with the temperature. It is 103.3 Ω^{-1} cm^{-1} at 1000 °C. $NiFe_2O_4$ nano-cermet displays high-temperature semiconductor properties.
4. The corrosion rate of $NiFe_2O_4$ nano-cermet in molten aluminum electrolyte is less than 3 cm/year, which can be used as inert anode for aluminum electrolysis.

The experimental results for the nano-cermet show some useful properties. The microstructure, mechanical properties, and the relationship between structure and performance should be analyzed, and the performance of nano-cermet should be improved.

References

1. Hanneman, R.E., et al.: Inert Anode Technologies Report. Report CRTD, vol. 53. American Society of Mechanical Engineers, New York, NY (1999)
2. Christini, R.A., Dawless, R.K., Ray, S.P.: Advanced Anodes and Cathodes Utilized in Energy Efficient Aluminium Production Cells. Final Technical Progress Report. Cooperative Agreement Number DE-FC07-98ID13666 (2001)
3. Edwards, L., Kvande, H.: Inert anodes and other technology changes in the aluminium industry—the benefits, challenges, and impact on present technology. JOM. **53**(5), 28 (2001)
4. Chinalco: Annual report (2016)
5. Ray, S.P., Liu, X., Weirauch, D.A.: Inert Anode Containing Metal Oxides, Copper and Noble Metal. US Patent 6,126,799 (2000)
6. Ray, S.P., Liu, X., Weirauch, D.A.: Cermet Inert Anode Used in Electrolytic Preparation of Metals. Chinese Patent 00,814,882.1 (2000)
7. Weyand, J.D., DeYoung, D.H., Ray, S.P., et al.: Inert Anodes for Aluminium Smelting. Aluminum Company of America, Washington, DC (1986)
8. Ray, S.P.: Inert anode for Hall cells. In: Miller, R.E. (ed.) Light Metals 1986, pp. 287–298. TMS, Warrendale, PA (1986)
9. Alcom, T.R., Tabereaux, A.T., Richards, N.E., et al.: Operational results of pilot cell test with cermet inert anodes. In: Das, S.K. (ed.) Light Metals 1993, pp. 433–443. TMS, Warrendale, PA (1993)
10. Kozarek, K.I., Ray, S.P., Dawless, R.K., et al.: Corrosion of Cermet Anodes During Low Temperature Electrolysis of Alumina. DE-FC07-89ID12848. Alcoa Inc, Lancaster, PA (1997)
11. Kozarek, K., Ray, S.P., Dawless, R.K., et al.: Energy Efficient Aluminum Production - Pilot-Scale Cell Tests: Final Report for Phase I and Phase E. DE-FC07-941D132I9. Alcoa Inc, Lancaster, PA (1999)
12. Christini, R.A., Dawless, R.K., Ray, S.P., et al.: Phase III Advanced Anodes and Cathodes Utilized in Energy Efficient Aluminum Production Cells. DE-FC07-98ID13666. Alcoa Inc, Lancaster, PA (2001)

13. Ray, S.P.: Electrolytic Production of High Purity Aluminium Using Ceramic Inert Anodes. US Patent 6,416,649, (2001)
14. Blinov, V., Polyakov, P.: Behaviour of cermet inert anodes for aluminium electrolysis in a low temperature electrolyte. In: Haarberg, G.M. (ed.) 11th International Aluminium Symposium, Norway, September 19–22, pp. 123–131 (2001)
15. Lorentsen, O.A., Thonsiadl, J.: Laboratory cell design considerations and behaviour of inert anodes in cryolite-alumina melts. In: Haarberg, G.M. (ed.) 11th International Aluminium Symposium, Norway, September 19–22, pp. 145–154 (2001)
16. Blinov, V., Polyakov, P., Krasnoyarsk, J., et al.: Behaviour of inert anodes for aluminium electrolysis in a low temperature electrolyte, Part I. Aluminium. **73**(12), 906–910 (1997)
17. Dawless, R.K., et al.: Reduced Temperature Aluminium Production in an Electrolytic Cell Having Inert Anodes. US Patent 6,030,518, (1997)
18. Office of Industrial Technologies: Advanced Anodes and Cathodes Utilized in Energy-Efficient Aluminium Production Cells. Aluminum Project fact Sheet, p. 2 (1999)
19. Belda, A.: Alcoa Quarterly Financial Report to Analysts (2001)
20. Wu, X., Wei, L., Jingmei, L., et al.: Nanoceramet TBR inert anode of aluminium electrolysis. In: Proc of the 2nd Int. Symp. on Met. & Mat. of Non-Ferrous Metals and Alloys, Shenyang China, 14–18 September (2004)
21. Qun, Z., Zhuxian, Q.: The ultimate challenge of high performance ceramic industry -- inert anode in aluminum electrolysis cells. Rare Metalmat. Eng. (9), 528–531 (2002) (in Chinese)
22. Qingwei, Q., Yanqing, L., Gang, Z.: Solid state synthesis of Ni-Zn ferrite for inert anode of aluminum electrolysis. Chin. J. Nonferr. Met. **13**(3), 769–773 (2003)
23. Kvand, H.: Inert electrodes in aluminum electrolysis cells. In: Light Metals 1999, p. 63. TMS, Warrendale, PA (1999)
24. Yexiang, L.: Progress in research and development of inert anode and wettable cathode for aluminum electrolysis. In: Light Metals 2001, p. 5. TMS, Warrendale, PA (2001) (in Chinese)
25. Gao, L., Wei, L.: Nano-ceramic, vol. 3, pp. 4–6. Chemical Industry Publishing House, Beijing (2002) 155–156
26. Lawn, B.R., Marshall, D.B.: Hardness, toughness, and brittleness: an indentation analysis. J. Am. Ceram. Soc. **62**, 347–350 (1979)
27. Boccaccin, A.R.: Machinability and brittleness of glass-ceramics. J. Mater. Process. Technol. **65**, 304–306 (1997)
28. Jee, J., Hwang, J.J., Mason, T.O., et al.: J. Mater. Res. **10**(9), 2295–2300 (1995)
29. Zhang, D.L.: Nanophase Materials, pp. 67–68. Chemical Industry Press, Beijing (2001)
30. Zhang, J., Chen, B.: Precision Processing Ceramics, vol. 6, p. 21. Shanghai Building Materials, Shanghai (2001)
31. Niihara, K.: New design concept of structural ceramics-ceramics nanocomposites. J. Ceram. Soc. Jpn. **99**, 974 (1991)
32. Niiharaa, K., Nakahiraa, A., Sekinoa, T., et al.: New nanocomposite structural ceramics. In: MRS Fall Meeting Pittsburgh, pp. 405–412 (1992)
33. Lian, G., Wei, L.: Nano Ceramics, vol. 3. Chemical Industry Press, Beijing (2002) (in Chinese)
34. Yexiang, L.: Research and development of inert anode and wet cathode for aluminum electrolysis. In: Light Metals, vol. 5, pp. 26–29. TMS, Warrendale, PA (2001) (in Chinese)
35. Rays, P.: Inert anodes for Hall cells. In: Light Metals 1986, pp. 281–298. TMS, Warrendale, PA (1986)
36. Alcoa: Test News, Pittsburgh (2000)
37. Zhuxian, Q.: Research and development of inert cathode and anode in aluminium electrolysis. In: Light Metals, vol. 9, pp. 30–34. TMS, Warrendale, PA (2001) (in Chinese)
38. Tabereaux, A.T.: Anode effects, PFCs, global warming, and the aluminium industry. JOM. **46**(11), 30–34 (1994)
39. Halvor, K.: Inert electrodes of aluminium electrolysis cells. In: Light Metals 1999, pp. 369–376. TMS, Warrendale, PA (1999)

40. Alcorn, T.R., Tabereaux, A.T., et al.: Operational result of pilot cell test with cermet "inert" anodes. In: Annual Meeting of the Metallurgical Society, Denver, CO (United States), pp. 21–25 (1993)
41. Windisch, C.F., Strascan, D.M., et al.: Materials characterization of cermet anode tested in a pilot cell. Mat. Sci, 445–454 (1993)
42. Wu, X., Wei, L., Licheng, Z., et al.: The Study on the Machine of Cermet for Inert Anode in Aluminum Electrolysis, vol. 12, pp. 25–28 (2007) (in Chinese)
43. Wu, X., Lina, A., Xiaohao, M.: Corrosion resistance research on inert anode of aluminium electrolysis. In: Light Metals, vol. 1, pp. 35–37. TMS, Warrendale, PA (2003)
44. Kechao, Z., Yuqiang, T.: Research development of nickel ferrite based cermet inert anode materials. Chin. J. Nonferr. Met. **21**(10), 2418–2429 (2011)

Chapter 5
Measurement of Mechanical Properties for $NiFe_2O_4$ Nano-cermet

5.1 Preparation of Nano-cermet Inert Anode Samples

To measure the mechanical properties of nano-cermet inert anodes, we reprepared the nano-cermet samples according to the preparation method of nano-cermet inert anodes (Sect. 4.4). Through the previous laboratory research, a set of relatively perfect preparation process of nano-NiO-Cu-$NiFe_2O_4$ cermet inert anode was basically formed, and the electrical conductivity and corrosion resistance of the prepared samples were tested. It is not the addition of nano-components but a relatively low mass percentage that can improve the brittleness of cermet materials. As reported by literatures [1, 2] and previous experiments, it is better to add nano-components in the range of about 8–10%. In this experiment, the ratio of raw materials is shown in Table 5.1.

In Table 5.1, each type of powder was put into a 100 mL small beaker, the proper amount of anhydrous ethanol was added as dispersant, and the ultrasonic wave was used to oscillate for 30 min. Then, the matter in each small beaker was merged into a large beaker of 500 mL, and the ultrasonic oscillation was continued for 30 min. After stopping, the pre-prepared sodium polyacrylate was added to the beaker to make the PH value of the slurry greater than 8. Then, the slurry was mixed into container of the QM-4H ball mill for ball milling. After ball milling, the slurry was removed, HCl was added to adjust the PH value less than 6 (Static 1 h), and then the upper layer of the liquid was removed. It was put into the oven for 14 h, the temperature is 100 °C, and then the compaction was made. The preparation process is shown in Fig. 4.4.

Because the sintering is a complex change process, the reasonable setting and control of the sintering molding process (such as temperature rise, constant temperature, heating cooling rate, and time) are the important guarantees of sample quality. The sintering temperature of the sample was determined to be 1100 °C, according to the previous experimental result.

W. Xianxi, *Inert Anodes for Aluminum Electrolysis*, The Minerals, Metals & Materials Series, https://doi.org/10.1007/978-3-030-28913-3_5

Table 5.1 The material quality of samples

Label raw material	A	B	C	D	E
Fe_2O_3 (g)	20.00	20.00	20.15	20.04	20.08
Cu (g)	8.50	8.50	8.50	8.60	8.56
NiO (g)	28.20	27.60	27.20	26.40	25.82
Nano-NiO (g)	1.80	2.40	3.58	3.80	4.28
Nano-NiO/NiO (the total) (%)	6	8	10	12	14
Atolin (g)	2.51	1.90	2.23	2.53	2.76

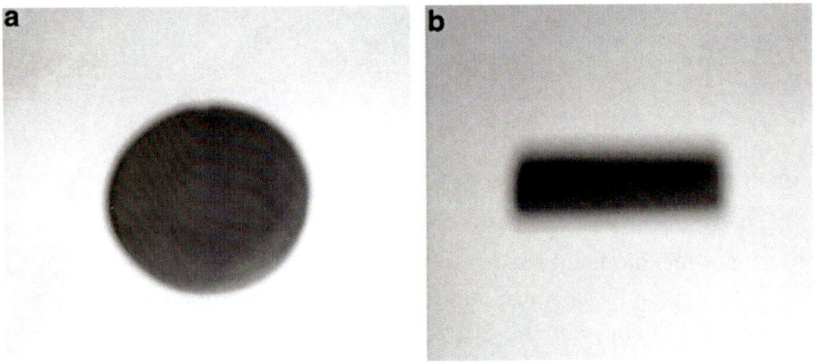

Fig. 5.1 The shape of sample before and after processing. (**a**) Sample shape before processing φ 30 (**b**) Processed sample shape 25 mm × 5 mm × 2.5 mm

5.1.1 Later Processing of Samples

After sintering, the samples of $NiFe_2O_4$-based nano-cermet were silver black with obvious metallic gloss. The reflection of crystal particles could be seen at the fracture surface after crushing, and the distribution of crystal particles was uniform. However, due to the shape of the mold, the sintered sample is round with φ of 30 mm in diameter, as shown in Fig. 5.1a. The next step of the mechanical property test for the sample is required to be strip. Thus, we need to process the sample.

5.1.1.1 Coarse Grinding

The rough grinding of a sample is carried out on the grinder by holding the specimen firmly and grinding it with the side of the grinding wheel. Due to the effect of grinding force, the grinding quantity of the upper part of the sample surface is too large. Thus, manual adjustment is required to make the force on the entire surface uniform.

5.1.1.2 Fine Grinding

After rough grinding, there are still thick and deep abrasions on the abrasive surface. In order to eliminate these feet, marks must be fine grinding. Fine grinding can be classified into manual grinding and mechanical grinding. This experiment uses manual grinding, metallographic sandpaper from coarse to fine (400-600-800-1000 mesh) for grinding. The sandpaper is spread on the glass board for one-way grinding, and the force on the sample is uniform. In this way, the whole grinding surface can be ground.

5.1.1.3 Polishing

The purpose of polishing is to remove the fine abrasions left over from the grinding surface and to obtain the bright and spotless mirror surface. In the experiment, the sample is polished on the polishing machine, and the switch is activated to turn the disk counterclockwise. The proper amount of polishing liquid drops on the disk can be polished.

5.2 Structural Characterization of Nano-cermet Specimen

After completing the above steps, the $NiFe_2O_4$-based nanocrystalline cermet samples were processed into strips with shape of 25 mm × 5 mm × 2.5 mm, as shown in Fig. 5.1b. The samples were silver black with obvious metallic gloss. The reflection of crystal particles can be seen on the surface of the sample. In order to ensure the reliability of the measured data, the fabricated cermet samples are characterized to ensure that the samples are $NiFe_2O_4$-based nano-cermet.
The *D*/max-*B* automatic diffractometer produced by Ricoh Optical Motor Company of Japan was used to test the experiment, and the program was automatically controlled by a microcomputer.

5.2.1 X-ray Diffraction Analysis Principle

The basic principle of X-ray diffraction technique is that a beam of electrons is accelerated at high pressure to bombard a metal target. A high-energy electron excites the inner shell electron (K layer electron) of the atom in the metal target. The electrons in the shell will move to the same orbit as the K electrons and emit characteristic X-rays. After the filtered X-ray is irradiated on the sample, the diffraction will occur when the wavelength λ of the X-ray is close to the crystal plane spacing of the sample. If the Bragg coherence condition is satisfied:

$$2d\sin\theta = n\lambda \tag{5.1}$$

The peak position of the diffraction peak is determined by the maximum interference. The intensity of diffraction peaks is determined by the different scattering factors and crystal structures of different atoms. The crystal structure, lattice parameters, and phase of the sample can be determined by the position of the diffraction peak (2θ) and the diffraction pattern.

5.2.2 Analysis of X-ray Diffraction

The principle of determining the grain size by X-ray diffraction is [3] that there is a basic parameter between the X-ray diffraction peaks of the same crystal phase to determine the grain size. Therefore, as long as more than two diffraction peaks of the same phase are selected, the grain size can be calculated by TOPASP software, according to the angle position, peak height, shape, and width of the diffraction peak. In the basic parameters, the relations between the physical parameters are preserved, and some secondary parameters are ignored. In addition, the basic parameters can distinguish the influence of the instrument on the diffraction peak. Thus, the grain size can be determined without the standard sample of non-fine processing.

According to Scherrer formula, the grain size D perpendicular to the direction of crystal plane is calculated by using the diffraction peak half maximum width $\beta_{1/2}$. If the diffraction peak width is only caused by grain refinement and the grain size is uniform, then the diffraction peak width is consistent with the Scherrer formula as follows:

$$D(h,k,l) = \frac{k\lambda}{\beta_{1/2}\cos\theta} \tag{5.2}$$

where D is the grain size along the perpendicular direction of crystal plane, k is the Scherrer constant with a value of 0.89, $\beta_{1/2}$ is the half maximum width of the diffraction peak, and λ is the wavelength of 0.15406 nm for the copper target used in X diffraction.

From the three strong lines of the characteristic peak of NiO in the diffraction diagram, the corresponding half-peak width and diffraction angle are found by looking up the PDF card, and then the grain size is calculated according to the Scherrer formula (Figs. 5.2, 5.3, and 5.4, Tables 5.2, 5.3, and 5.4).

From the phase analysis of diffraction patterns, we found that most of the materials obtained are $NiFe_2O_4$, NiO, and $Cu_{0.2}Ni_{0.8}O$ in addition to some oxidized CuO, indicating that the target product $NiFe_2O_4$ is formed. The composition of the sample is relatively simple.

According to the above principles and formulas, the average size of nanocrystalline NiO grains is about 20–30 nm.

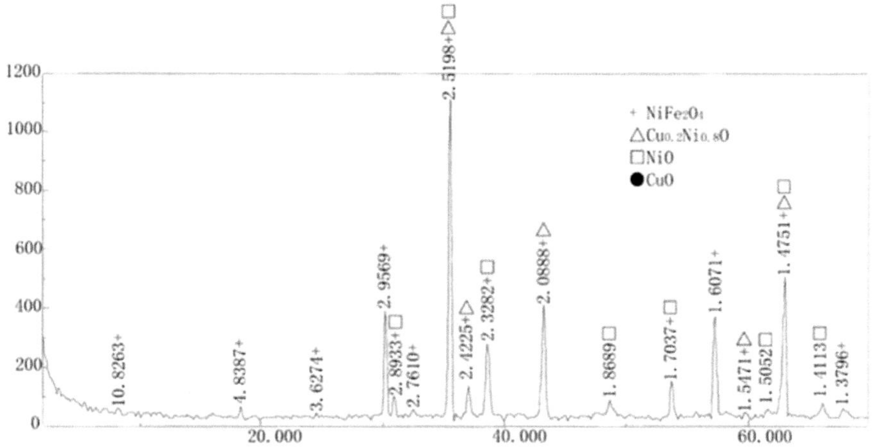

Fig. 5.2 The XRD atlas of sample for 6% nano-NiO

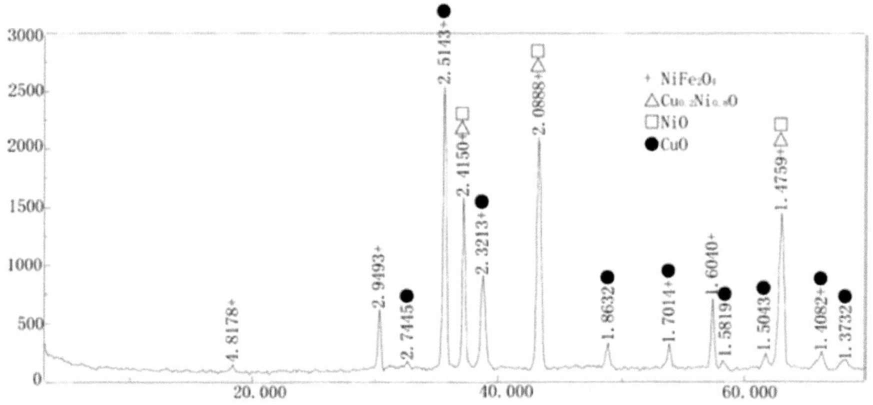

Fig. 5.3 The XRD atlas of sample for 10% nano-NiO

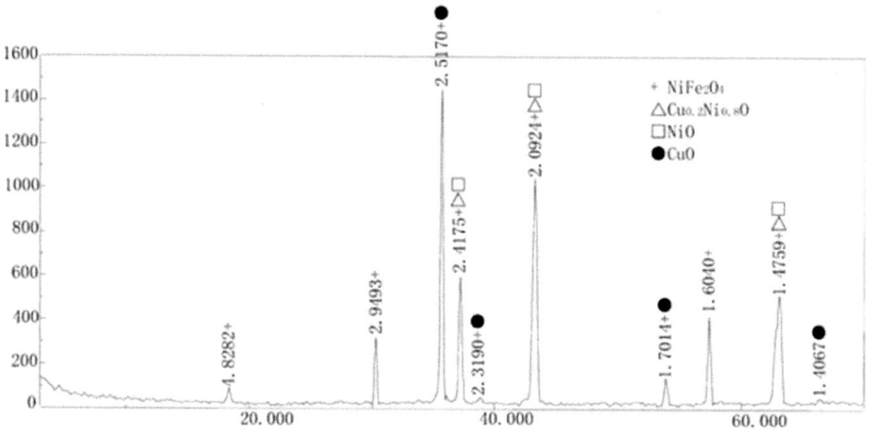

Fig. 5.4 XRD picture of 12% nano-NiO sample

Table 5.2 The strength, half the peak width, and grain size of the sample with 6% nano-NiO

Number	Intensity (I)	Peak width at half height ($\beta_{1/2}$)	Angle of diffraction (θ)	Crystal size (D/nm)
1	2.4150	0.00574	18.60	25.20
2	2.0906	0.00812	21.62	17.96
3	1.4768	0.00272	31.44	31.81

Table 5.3 The strength, half the peak width, and grain size of the sample with 6% Nano-NiO

Number	Intensity (I)	Peak width at half height ($\beta_{1/2}$)	Angle of diffraction (θ)	Crystal size (D/nm)
1	2.4075	0.00539	18.66	25.21
2	2.0833	0.00742	21.70	19.98
3	1.4751	0.00269	31.48	32.05

Table 5.4 The strength, the half peak width, and the grain size of the sample with 12% NiO

Number	Intensity (I)	Peak width at half height ($\beta_{1/2}$)	Angle of diffraction (θ)
1	0.00471	21.64	31.32 nm
2	0.00518	18.12	27.93 nm
3	0.00518	31.5	31.04 nm

According to the definition of nanocomposite ceramics, nanometer $NiFe_2O_4$ cermet were fabricated experimentally.

The other samples were tested in the same method, and the results were confirmed as nano-$NiFe_2O_4$ cermet.

5.3 Testing of Mechanical Properties of $NiFe_2O_4$ Nano-cermet

The mechanical performance testing of material plays an important role in evaluating the development and application of materials, especially in material design [4]. In order to design and process high-performance cermet, it is necessary to fully grasp the mechanical properties of cermet, so as to clarify the difference between the development goal and the used material. In addition, we determined the development policy of high-performance cermet. In addition, to design cermet reasonably and accurately according to the operating conditions and environment, reliable and real testing and evaluation methods are needed. Therefore, the determination of testing and evaluating methods for mechanical properties of cermet is an important research subject for correctly designing cermet, ensuring mechanical properties and quality of use, and expanding the scope of application. The test of mechanical properties is very important for analyzing the effect of nano-component NiO materials on the overall properties of $NiFe_2O_4$ cermet. The mechanical properties of the samples were tested by hardness, flexural strength, and fracture toughness.

5.3.1 Hardness Test of $NiFe_2O_4$ Nano-cermet

Hardness measurement is one of the most widely used methods for evaluating the mechanical properties of materials. It is easy to obtain data and plays an important role [5]. Hardness is one of the important mechanical properties of materials and is the characterization of deformation ability of materials to resist local pressure. Hardness refers to the ability of a material to resist deformation or fracture in a small volume on the surface. It is a comprehensive parameter to characterize the properties of materials. The hardness of the material can be regarded as an ability to resist plastic deformation on the surface of the material under the action of contact stress. Hardness measurement can give a quantitative concept of the hardness and softness of the material. The stress and the degree of deformation of the material at different depths below the surface of the material are different. Thus, the hardness value can reflect the elasticity of the local volume near the indentation, the micro-plastic resistance, the plastic deformation strengthening ability, and the large amount of deformation resistance. The higher the hardness value, the greater the resistance of the material to plastic deformation, and the more difficult it is for the material to produce plastic deformation.

5.3.1.1 Hardness Measurement Method

There are many methods to determine the hardness, usually classified into two categories: engraving method and pressing method. In the production, the pressing method is more commonly used, including Brinell hardness, Rockwell hardness, Vickers hardness, and microhardness. The physical meaning of hardness refers to the plastic deformation caused by the resistance of the material surface to local compression of a harder object than itself.

The common methods of measuring the hardness of ceramic matrix composites are macro and micro Vickers hardness. They all choose proper load and use special mold head to press out indentation on the polished surface of the material. The main difference between them is that the load is different, so the size of the indentation produced is also different. The microhardness is usually applied between dozens of grams and hundreds of grams, and the indentation is observed and measured through a microscope. In this experiment, the microhardness method was used to meet the requirements of the specimen.

5.3.1.2 Principle of Micro Vickers Hardness Measurement

The general principle of hardness measurement is to apply a certain proportion of load at a certain time interval to press the hard indenter into the surface of the material to be measured. After removing the indenter, the depth or area of the indentation is measured. The Vickers hardness test is based on the indentation unit area of the

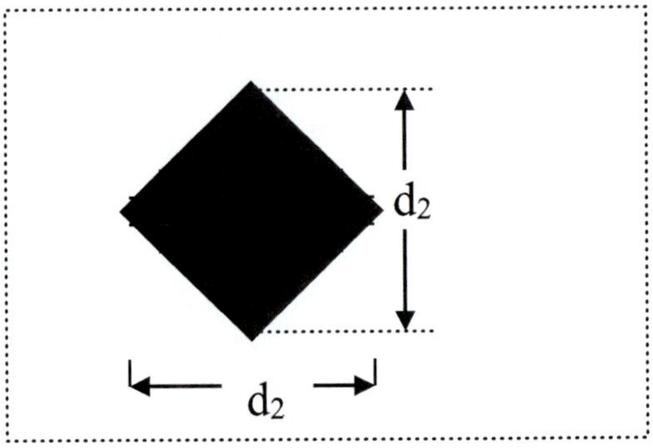

Fig. 5.5 The map for the principle of Vickers hardness

test force to express the Vickers hardness value. Under a certain case, the load is F (in kgf or N). We pressed the diamond tetrahedral cone with two pairs of angles 136° into the surface of the tested cermet (as shown in Fig. 5.5) and pressed a quadrilateral cone indentation on the surface of the specimen. Then we removed the load, and the average diagonal length d of indentation is measured with a reading microscope.

$$d = \left[(d_1 + d_2) / 2 \right] \tag{5.3}$$

where d_1 and d_2 are diagonal lengths in two different directions to calculate the surface area of indentation. Therefore, Vickers hardness value (HV) is the quotient of test force P divided by indentation surface area F. When the test force P is kgf (9.8 N), the formula is as follows:

$$\text{HV} = \frac{P}{F} = \frac{2P\sin\left(136^{\circ}/2\right)}{d^2} = 1.8544\frac{P}{d^2} \tag{5.4}$$

When the test force P is N, the formula is as follows:

$$\text{HV} = \frac{0.102P}{F} = \frac{0.204P\sin\left(136^{\circ}/2\right)}{d^2} = 0.1891\frac{P}{d^2} \tag{5.5}$$

5.3.1.3 Experimental Procedure

In this study, the HvS-1000 Vickers hardness tester was used to measure the hardness of the samples. The specific steps are as follows.

After cutting the samples, the samples were roughened and then were polished on the PG-2B metallographic polishing machine, and the grinding agent was diamond grinding paste. We polished the coarse sample until there is no scratch on the surface of the sample observed by microscope. Finally, five indentations were made on the surface of the sample with diamond indenter on the hardness meter. The

Table 5.5 The data of microhardness sample test

Label	A1	A2	A3	A4	A5	A6
Nano-NiO (vol%)	0	6	8	10	12	14
HV (average value)	1465	525	438	391	476	422
Standard deviations	92.5	20.4	23.6	27.5	36.4	19.8

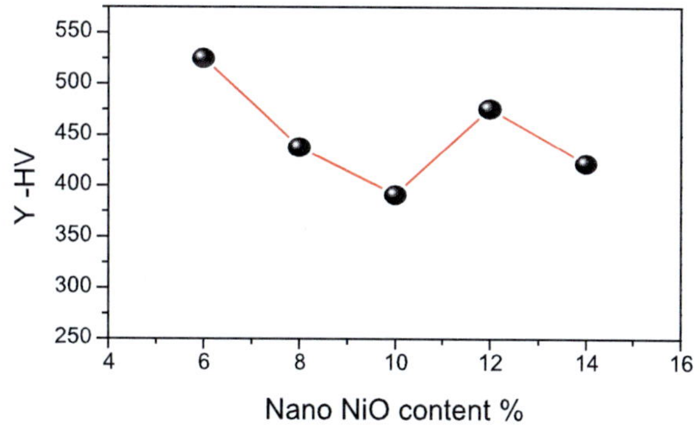

Fig. 5.6 The relationship between sample hardness and nano-NiO content

pressure was 1 kgf, and the holding time was 15 s. After removing the load, the length of the two diagonal lines of the square cone indentation is accurately measured with the optical microscope attached to the hardness meter. The average length of the two diagonal lines and the above Formula (5.5) are used to calculate the average hardness of the specimen.

5.3.1.4 Experimental Result

After indentation, the HvS-1000-type Vickers hardness tester was used in this experiment. Because it is unnecessary to add the Formula (5.5) in the instrument design to the calculation, we directly measured the diagonal length to read the hardness value. The hardness value of the sample is as shown in Table 5.5.

According to Table 5.5, the hardness of the samples without nano-NiO is much higher than that of the samples containing nano-NiO, showing the high-hardness properties of ceramic materials. Therefore, the addition of nanometer NiO has a great influence on hardness. However, the hardness of the samples containing nano-NiO first decreases and then increases as the increase of nano content, as shown in Fig. 5.6.

As shown in Fig. 5.6, at first, the hardness value of 6% nano-NiO content is higher, and the inflexion point reaches the minimum value of 391 HV at 10% nano-NiO content. The results show that the hardness of 10% nanocrystalline is the best (minimum).

5.3.2 *Bending Strength Test*

Strength is one of the most important mechanical properties of structural materials. Structural materials are often subjected to heavy loads during use. Strength is a kind of ability index to resist the damage when the material is under stress. It refers to the maximum stress that the material can bear, that is, the stress when the material is loaded to fracture. Strength, like hardness, is an important physical parameter. It is determined by the composition and structure of the material and changes with the change of external conditions (such as temperature, stress state, etc.). The actual strength of ceramic material is much lower than theoretical one because of its various defects. Usually, the strength of ceramic materials is expressed by flexural strength. The bending strength test of ceramic materials has been widely used in the test of ceramic materials because of its simple processing and convenient operation. The flexural strength test is the main experimental method for evaluating the strength of ceramic materials. This method is simple and feasible and can be compared among different materials.

The flexural strength of the material is affected by the composition of the material, the microstructure, and the defect of the material. The porosity, inclusion, grain boundary, density, crack, residual stress, and other defects in the ceramic material lead to the wide distribution range of the strength of the ceramic material. The bending strength of the material is closely related to the machinability of the material. Thus, it is necessary to analyze the flexural strength of the material.

5.3.2.1 Three-Point Bending Test Principle

Generally, ceramic materials are brittle. Except for special requirements, the tensile strength test is rarely conducted, usually the bending strength test. The most common method used in bending strength testing is the three-point bending test, which is simple and comparable among different materials [6]. The principles of the test are as follows.

When a beam with a rectangular section is subjected to bending load, stress occurs on the section (Fig. 5.7). The stress can be divided into normal stress perpendicular to the section and shear stress parallel to the section. If the loads on the beam are in the same plane and perpendicular to the central axis of the beam, the normal stress at each point of the section is combined into a couple, and its moment is known as the bending moment *M*.

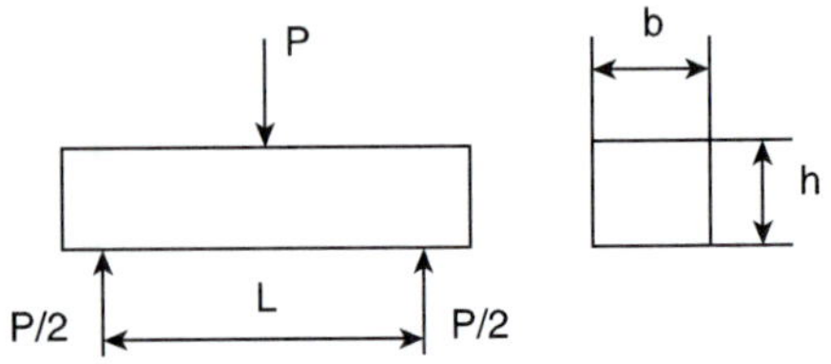

Fig. 5.7 The cermet bending strength of three-point test

The normal stress at any given point is proportional to the moment from that point to the middle section and the moment on the section and inversely proportional to the moment of inertia of the section to the middle section. If the bending moment on the section is positive, the compressive stress is above the point of the middle section, and the tension stress is at the point below the middle section; if the bending moment on the section is negative, the situation is opposite. Therefore, the maximum stress expression on the external surface of the tension side of the beam is $\sigma_{max} = M_{max}/W$. The maximum bending moment $M_{max} = PL/4$ for three-point bending is at the midpoint of beam length, as shown in Fig. 5.7. Due to the rectangular section of the section moment $W = bh^2/6$, the expression of bending strength is:

$$\sigma = \frac{3Pl}{2bh^2} \tag{5.6}$$

where σ is the flexural strength (MPa); P is the load added (N); l is two fulcrum span (mm); b is the width of specimen (mm); and h is the height of specimen (mm).

5.3.2.2 Test Step

The instrument used in the experiment is WDW-10C electronic universal material machine with a span of 25 mm and a loading rate of 0.5 mm/min.

1. Preparation of the sample: the size of sample is approximately 25 mm × 2.5 mm × 5 mm after the sample is polished.
2. Measurement of specimen size: measuring and recording the width of the specimen with Vernier caliper to 0.01 mm.
3. Sample placement: the specimen is placed on a special mold, and the pressure head is adjusted to be close to the sample, leaving a gap of about 1.0 mm.
4. Automatic measurement: on the computer control end of the material experiment machine, the maximum load force is 250 N, and the loading rate is 0.05 mm/min, and then the "start" button is clicked for automatic measurement.
5. Repeated test: because of the statistical distribution of the strength of the material, in order to objectively reflect the strength of the material, the more samples must be tested. Then, the average value is closer to the real value. However, it is not allowed to test too many samples from an economic point of view. In this experiment, five samples of each material are selected for repeated test.

5.3.2.3 Experimental Results and Analysis

According to the data in Table 5.6, the average flexural strength of the samples added nano-NiO is greater than that of the samples without NiO. The addition of nano-NiO can improve the flexural strength of the samples. The results show that the addition of nanometer NiO has a good effect on bending strength.

As shown in Fig. 5.8, the flexural strength of the sample with the addition of 6% nano-NiO is higher than that of the sample without nano-NiO. However, the increase is not obvious. The effect of nano-NiO is not significant.

Table 5.6 The data of experimental bending strength of specimens

Number	A0	A1	A2	A3	A4	A5
Nano-NiO content %	0	6	8	10	12	14
Bending strength (MPa)	75.78	84.16	108.80	142.41	113.10	98.41
Standard deviation	3.81	1.13	2.54	3.62	4.25	1.35

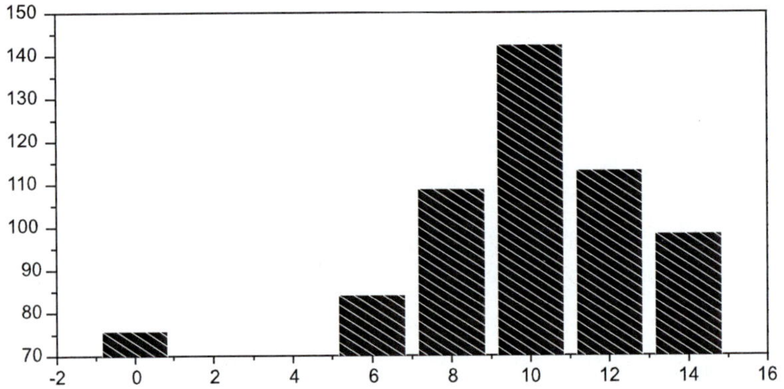

Fig. 5.8 The relationship between flexural strength and nano-NiO content of sample. *x*—Nano-NiO content, *y*—Flexural strength

According to Gurland strength theory [7], the relationship between the strength and the structural parameters of the material conforms to the law of dispersion strengthening. The critical fracture strength can be estimated by the strain energy required for the propagation of crack by nanoparticles, and the critical stress expression can be obtained:

$$\sigma^2 = K'\varphi^{2/3} / d \tag{5.7}$$

where σ is the axial stress; φ is the volume fraction of NiO; d is the average particle size of NiO; and K' is a constant.

According to the above formula, the smaller the size of nano-NiO, the higher the bending strength. The flexural strength of the sample increases greatly when the content of nano-NiO increases to 8%. When the content of nano-NiO reaches 10%, the flexural strength of the sample reaches a maximum value of 142.41 MPa.

However, when the content of nano-NiO is more than 10%, the grain size of nano-NiO increases and the flexural strength of the sample decreases due to the influence of sintering. Nano-NiO can refine the grain and prevent the abnormal growth of the crystal.

As shown in Fig. 5.9, the curve of A0 sample is steep, the displacement of A0 sample is nearly 0.2 mm more than that of A3 sample, and the curve trend of A3 sample is moderate. The peak value of test force of A3 sample (124.17 N) is much greater than that of A0 sample (71.23 N), which indicates that the bending strength of A3 sample is significantly greater than that of A0 sample.

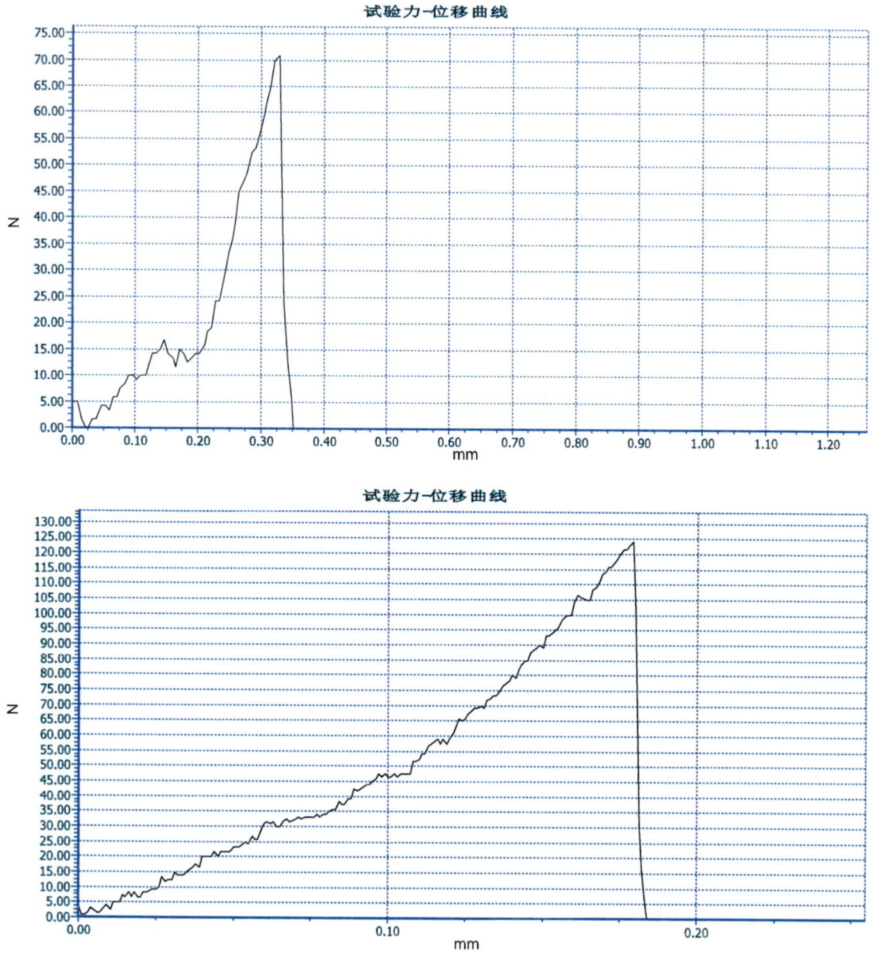

Fig. 5.9 The test force vs. displacement curve of sample A0 and A3 in bending strength test. (**a**) Sample A0 containing nano-NiO of 0%. (**b**) Sample A3 containing nano-NiO of 10%

5.3.3 Testing of Fracture Toughness

5.3.3.1 Significance of Fracture Toughness Testing

The fracture toughness is one of the important mechanical properties of materials. It is the inherent properties of materials and related to the microstructure of materials. It is also a characterization parameter for evaluating the resistance of materials to brittle failure or crack instability propagation. The brittle fracture of materials is the most dangerous one. It is often due to the existence of various internal and surface defects, such as cracks, notches, scratches, voids, and inclusions, which cause stress concentration. Based on the characteristics of cermet materials, fracture mechanical properties have become an important index to evaluate the mechanical properties of cermet materials. The most common mechanical parameter used to evaluate fracture toughness of cermet materials is fracture toughness. The failure of brittle materials is often catastrophic, that is, once the crack in the material is propagated, it will immediately reach the unstable state, and the crack will grow like a broken bamboo. The fracture toughness of the material can be used to measure the ability to resist crack propagation, that is, to resist brittle failure. It reflects the material plasticity and the inherent property of material. There are three types of crack propagation: break type, staggered type, and tear type. The break type is the most severe one, so it is usually used to measure the fracture toughness of the material. The measured value is called K_{IC}. The K_{IC} value of the material is not affected by the crack and geometric shape under the plane strain state. Thus, it is very important to understand the essential properties of cermet, a multi-crack material.

Fracture toughness is a method to study ceramic fracture from the perspective of crack surface displacement and crack tip stress. The study of fracture toughness (K_{IC}) not only contributes to the application of $NiFe_2O_4$ nano-cermet as inert anode material but also has positive significance for the development of new materials and the production process [8].

5.3.3.2 Fracture Toughness Testing Method

At present, there are a variety of testing methods for fracture toughness of cermet materials. The main testing methods for fracture toughness of cermet materials include indentation method [9, 10], single-edge notched beam (SENB) method, double torsion (DT) method, mountain notch split method, indentation method, indentation fracture method, and so on. Some methods have high technical difficulty and are not easy to be applied on a large scale, and some methods have great measurement errors and certain difficulties in their application. However, these methods have not been given special priority for a long time, because each of them has shortcomings, which limits their application [11]. Compared with the research material in this study, the indentation method and SENB method are mainly used. The indentation method has the advantages of small sample, simple operation, less time, etc. Thus, it has attracted more and more attention.

5.3.3.3 Vickers Indentation Method

Compared with the traditional methods, the indentation method has been paid more attention to because of its advantages of simple operation and low cost, especially in the rapid quality inspection of ceramic material production. Vickers indentation hardness is the most widely used indentation technique, which has been widely used for decades. This method is particularly useful for brittle materials with low fracture toughness. It has been widely used to measure fracture toughness of brittle materials such as ceramics [12, 13]. The test principle is that at the same time as the Vickers hardness test, the crack length produced by the indentation and its four corners is

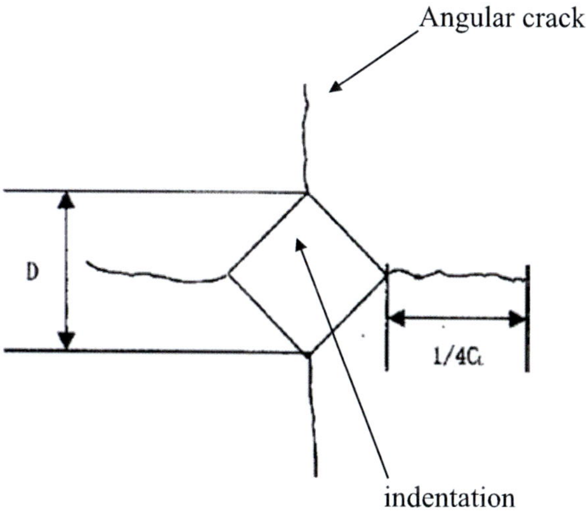

Fig. 5.10 Vickers indentation

shown in Fig. 5.10, and the Vickers hardness value HV, according to the formula (5.8) KIC obtained.

The formula for calculating fracture toughness is

$$K_{IC} = 0.113 \mathrm{HV} D^{1/2} \left(1 + C_L / 2D\right)^{-3/2} \tag{5.8}$$

where K_{IC} is fracture toughness (Pa $m^{1/2}$); HV is Vickers hardness (kg/mm^2); P is load (kg); D is the average length of diagonal indentation (mm); and C_L is the average total crack length (mm).

However, according to the theory of indentation fracture mechanics established by Lawn et al. [14], the only driving force of indentation crack formation and propagation is the residual stress caused by elastic-plastic deformation mismatch of materials near indentation. The driving force of cracks in Vickers indentation testing process is the residual stress caused by the mismatch between the plastic zone beneath the indentation and the elastic matrix around it. The results reflect the local crack propagation resistance of the material. It should not be the macroscopic fracture toughness of the material. This has great influence on the application of measuring fracture toughness of ceramic matrix composites.

5.3.3.4 Single-Edge Notched Beam Method

Single-edge notched beam (SENB) is the most common method for testing the toughness of ceramics. The sample processing is relatively simple, and it is suitable for testing at high temperature or in atmosphere. The data has low dispersion and good reproducibility. Compared with double torsion method or indentation method, the single-edge notched beam method is the most mature and widely used method for testing the fracture toughness of cermet. The single edge notch beam method

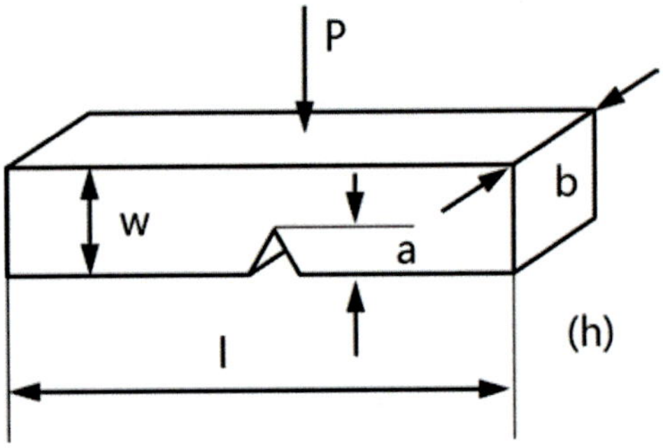

Fig. 5.11 The testing of fracture toughness by SENB method

Table 5.7 The value of coefficient *D*

Coefficient *D*	D_0	D_1	D_2	D_3	D_4
Value	2.9	−4.6	21.8	−37.6	38.7

reflects the macroscopic fracture toughness of the material [15]. If we want to obtain more accurate fracture toughness values to compare the macroscopic fracture toughness of materials, the indentation method cannot be selected. Thus, this experiment uses single-edge notched beam method.

The single-edge notched beam method is roughly the same as the bending strength test process. The difference is that the crack is prefabricated in the middle of the specimen, and the depth is about half of the height of the specimen, as shown in Fig. 5.11.

When the size ratio of the sample is $a/W = 0.4–0.6$, $L/W = 4$, and $b \approx W/2$, the fracture toughness formula is

$$K_{IC} = \sigma Y a^{1/2} = PLf\left(a/W\right)/bW^{3/2} \tag{5.9}$$

where $f(a/W) = D_0(a/W)^{1/2} + D_1(a/W)^{3/2} + D_2(a/W)^{5/2} + D_3(a/W)^{7/2} + D_4(a/W)^{9/2}$.

The value of coefficient *D* is shown in Table 5.7. Among them, *L* is the specimen span; *W* is specimen height; *a* is crack depth; *b* is specimen width; and *P* is fracture load.

5.3.4 *Experiment of Fracture Toughness*

5.3.4.1 Experimental Procedure

1. The specimen size is 25 mm × 2.5 mm × 5 mm after polishing. The crack is prefabricated in the middle of the specimen by internal circular cutting machine, and the crack depth is about half of the height.

2. The specimen is put into the special mold and then put into the WDW electronic material universal testing machine. The loading rate is 0.05 mm/min. After the specimen is broken, the ballast load is recorded.
3. Using Vernier caliper to measure and record its width b and height w. The precision is 0.01 mm.
4. Repeated test of five samples selected for each content.

5.3.5 *Experimental Results and Analysis*

The average value of fracture toughness of each sample is listed in Table 5.8. The relationship between fracture toughness and nano content of the sample is shown in Fig. 5.12.

The fracture toughness of the material has been improved due to the addition of nano-NiO, as shown in Fig. 5.12. The fracture toughness of the samples increases from 2.28 to 4.25 MPa $m^{1/2}$ as the content of nano-NiO increases from 0 to 12. The results show that the addition of nano-NiO has significant influence on toughening [16, 17].

Table 5.8 The fracture toughness specimen

Identifier	A0	A1	A2	A3	A4	A5
Nano-NiO content %	0	6	8	10	12	14
Fracture toughness K_{IC} (MPa $m^{1/2}$)	2.28	2.97	3.81	4.02	4.25	3.96
Standard deviation	0.09	0.21	0.15	0.08	0.14	0.14

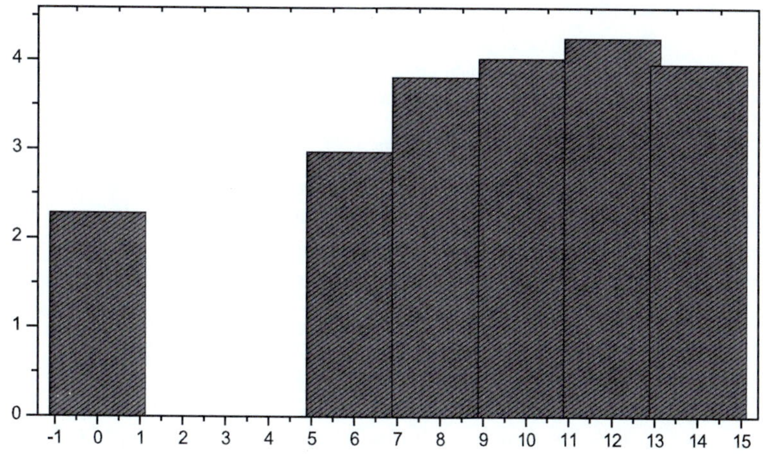

Fig. 5.12 The relationship between the nano-NiO content and fracture toughness. x—Nano-NiO content, y—Fracture toughness (MPa $m^{1/2}$)

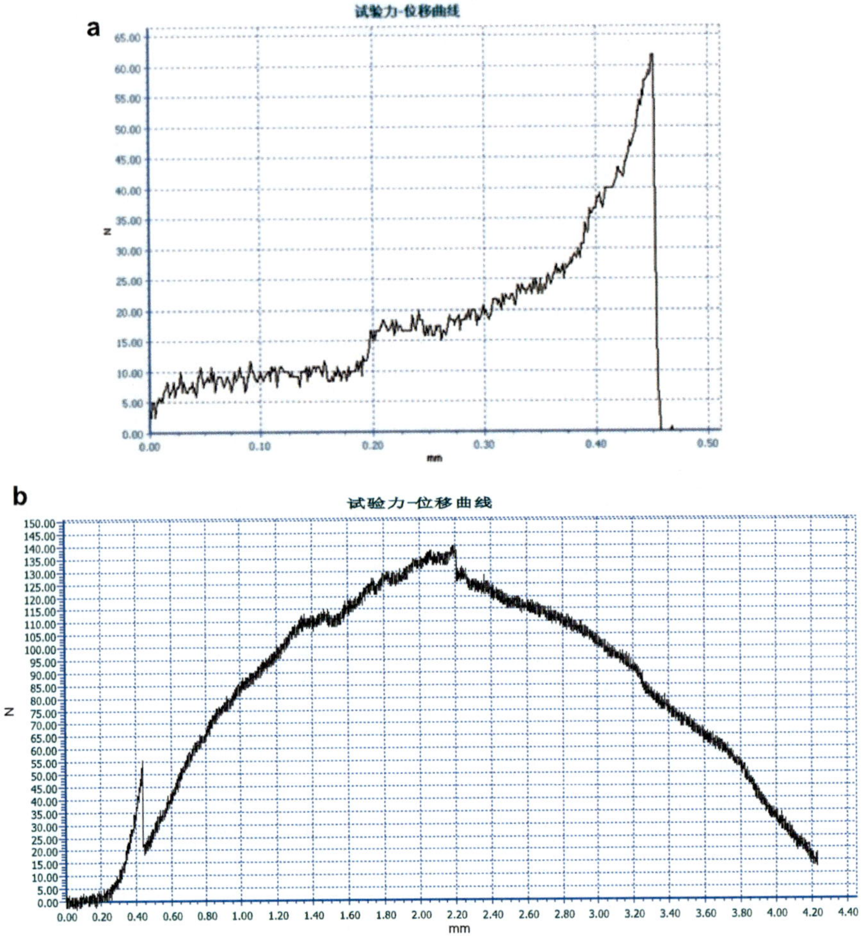

Fig. 5.13 The test force vs. displacement curve of specimens A0 and A4. (**a**) Sample containing nano-NiO of 0%, A0. (**b**) Sample containing nano-NiO of 12%, A4

As shown in Fig. 5.13a, the test force increases slowly when the sample does not add nano-NiO. When the maximum load is reached (61.67 N), the specimen has brittle fracture. The test force curve decreases linearly. The sample with nano-NiO of 12% is shown in Fig. 5.13b. At the displacement of 0.4 mm, the test force has a peak value of 55.78 N. It shows that the local fracture occurs along the crack in the experiment. With the increase of test force, the specimen remains resistant until the maximum load (139.64 N) is reached. After fracture, the curve of test force does not decrease linearly like the specimen without nano-NiO and slowly decreases with the deepening of the displacement. This shows that the addition of nano-NiO has a positive significance for toughening.

5.4 Microstructure Analysis of $NiFe_2O_4$ Nano-cermet

5.4.1 Toughening Mechanism of Nanocomposite Ceramics

Nanometer multiphase ceramics were first proposed by Roy et al. [18] in the preparation of ceramic metal multiphase materials by sol-gel method. The ceramic-based nanomaterials were first obtained by Professor Niihara in the study of Al_2O_3/SiC multiphase materials [19]. Although these four types of nanocomposites have different reinforcing and toughening mechanisms, all of them can significantly improve the mechanical properties, such as hardness, strength, creep resistance, and fatigue failure, and increase the reliability of ceramic materials in the process of application.

Improving the properties of materials by multiphase composite is one of the important methods of material modification. Most of the multiphase ceramics studied in the past are micron multiphase ceramics, that is, the grain size of each phase in this kind of material is in the range of micron. The second-phase particles are mainly distributed at grain boundaries, which are reinforced and toughened by stress-induced phase transition, microcrack toughening, crack deflection, pinning, and bridging. However, each phase or at least one of the phases in the novel nanocomposite ceramics is nanoscale in one dimension. Nanometer ceramics, as a new type of high-performance ceramics, became the most important new technology in ceramic research. This has attracted more and more attention of scientists all over the world. The research and development of nanoceramics will inevitably lead to the development and transformation of ceramic industry, as well as the development of ceramic theory and even the establishment of a new theoretical system. This can meet the needs of nanoscale research and make nanoceramic materials have better properties.

Nanocrystalline multiphase ceramics can be classified into three types: intragranular type, grain boundary type (the combination of these two types can form the intergranular grain boundary type, as shown in Fig. 4.2b), and nanocomposite type, as shown in Fig. 5.14.

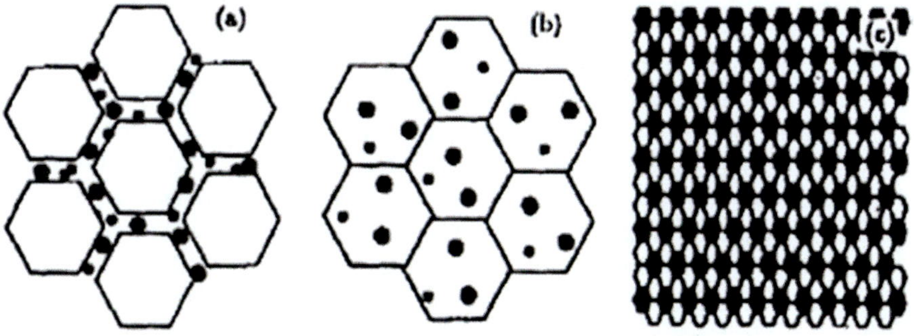

Fig. 5.14 (a–c) Structure of different nanocomposite

The second phase of nanoparticles is mainly distributed in the grain of matrix in intracrystalline nanocomposite ceramics, and the second phase of the grain boundary nanocomposite ceramics is mainly distributed on the grain boundary of the matrix. However, in nano/nano-multiphase ceramics, all phases are nanocrystalline. Some nano/nano-multiphase ceramics exhibit good processability or superplasticity due to their special nanostructures. The strength of nanocomposites is much higher than that of matrix materials and conventional micron composites. However, the key to the maturation of nanocomposites is to further improve the toughness of nanocomposites. In conventional micron composite ceramics, dispersion strengthening of hard particles, toughening of phase transition, and whisker compounding are common methods to improve the toughness of materials. When microparticles and nanocrystalline particles are introduced into the matrix at the same time, the microstructure will be formed as shown in the figure. The nanoparticles and micron particles in this kind of materials can play a synergistic role in toughening and reinforcing the matrix and greatly improve the properties of the materials. In addition, the self-reinforcing structure of the material can be obtained by controlling the shape of the matrix grain, and the better toughness can be obtained under the premise of ensuring the formation of the intragranular structure without causing cracks in the porcelain body. It is expected that nanocomposites with high strength and toughness can be obtained by strengthening the crack deflection and bridging action of nanoparticles.

The toughening mechanism of ceramic matrix composites varies, and it can be classified into two categories: "processing zone mechanism" and "bridging mechanism" [20]. The processing zone mechanism can be classified into two types: microcrack toughening and crack deflection toughening.

5.4.2 *Microcrack Toughening Mechanism*

The toughening mechanism of microcracks (microcracks mechanism) was proposed in the early 1970s [21]. The toughening forms of ZrO_2 can be classified into two main types. One is that the phase transition of ZrO_2 causes microcracks, that is, ZrO_2 will cause part of the volume expansion under the stress-induced phase transition, thereby hindering crack propagation to produce toughening effects [22]. The other one is that the addition of toughening phase causes the mismatch of thermal properties between the matrix and the second phase. The residual stress is produced during the material preparation from the high-temperature state to the room-temperature cooling, and it leads to the turning or branching of the propagation path during the crack propagation (Fig. 5.15). When the crack and the second phase meet in the multiphase ceramics, the propagation path of the main crack will be changed, resulting in bending, inclination, and distortion, forming the non-plane crack surface, increasing the fracture surface, consuming more crack propagation energy, and playing a toughening role.

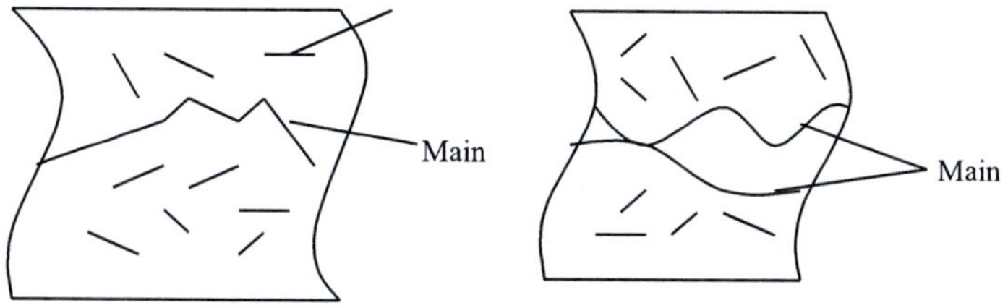

Fig. 5.15 The mechanism of microcrack increase of tenacity

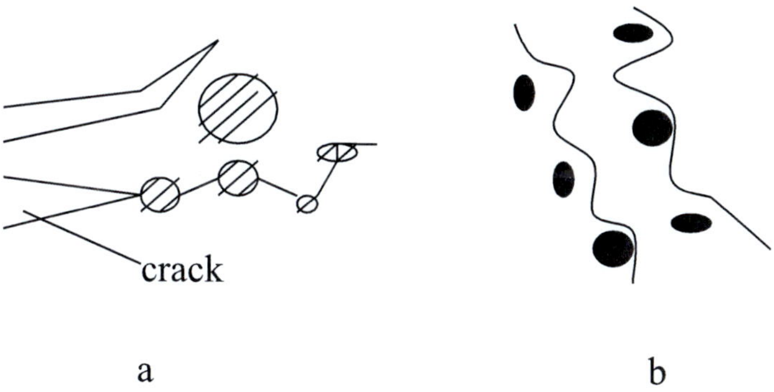

Fig. 5.16 Sketch map of the crack deflection and bend. (**a**) Crack deflection. (**b**) Crack bending

5.4.3 *Crack Deflection and Crack Bending Toughening Mechanism*

Crack deflection refers to the propagation of crack in the process of crack growth, in which the second-phase particle or residual stress field deviates from the original direction of motion and appears sawtooth growth in the material [23]. The crack bowing is due to the hindrance of strengthening phase to bend the crack tip path. In this case, the crack plane is reoriented perpendicular to the direction of applied tensile stress. The variation of this direction means that the crack propagation path will be increased, and the crack deflection will increase the toughness of the material because the crack plane is no longer perpendicular to the tensile stress direction and the stress intensity at the crack tip decreases, as shown in Fig. 5.16.

5.4.4 *Crack Bridging Toughening Mechanism*

Crack bridging toughening is a common toughening mechanism in multiphase ceramics. Crack bridging is an (crack wake) effect at the tip of a crack, when the crack tip is connected by a reinforcing agent to the two surfaces of the crack. This

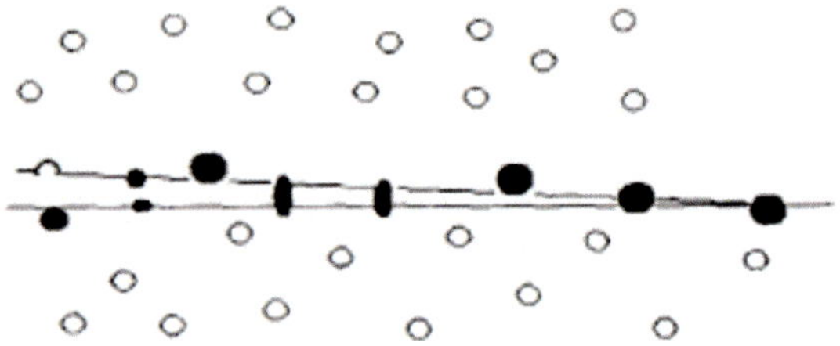

Fig. 5.17 Sketch map of the bridge toughening

provides a stress that makes the two surfaces of the crack close to each other, that is, the closed stress. As a result, the stress intensity factor increases with the crack propagation. Adding ductile particles to the brittle ceramic matrix can significantly improve the fracture toughness of the material. The mechanism of crack bridging toughening is the energy dissipation behavior occurring in a wide range after the crack tip. It can be divided into two forms. One is the rigid phase leading to crack bridging, and the other is ductile phase leading to crack bridging. When the second phase is a rigid phase and similar to the ceramic matrix, bridging toughening requires a certain aspect ratio of the second phase. Thus, they can be fibers or whiskers, columnar particles, and even basic particles with large aspect ratios. In addition, proper interfacial bonding states are needed. The weak interfacial bonding can be toughened by sliding, dissociating, and even pulling out the second phase to form a crack surface. The drawing effect in the bridging stage can greatly strengthen the crack bridging and toughening process. The ductile phase leads to the crack bridging, and the ductile particle refers to the metal particles. The elastic strain of metal particles makes crack bridging become the most effective toughening mechanism in cermet when the crack extends to the ceramic/metal interface [23]. Due to the different deformability of ductile metal particles and brittle matrix, the local passivation of cracks is caused, and some crack segments are forced to cross the particles and form elongated metal particles bridging [24], as shown in Fig. 5.17.

5.5 Observation and Analysis of Fracture Morphology of $NiFe_2O_4$ Nano-cermet

Compared with these conventional ceramics, the strength and toughness of nano-multiphase ceramics have been greatly improved. The toughening mechanism of nano-multiphase ceramics has always been the focus of attention, and the effect of nano-NiO on it is analyzed by electron microscope.

5.5.1 A Brief Introduction to the Structure and Working Principle of Scanning Electron Microscopy (SEM)

The main structure of scanning electron microscope is divided into five parts: electronic optics system, scanning system, signal receiving, amplifying, and displaying system, sample fretting and replacing system, and vacuum system.

Its working principle can be explained as follows. The electron beam emitted by the high-voltage filament power supply is focused on the grid, passes through two or three electromagnetic lenses under the action of an accelerating voltage, and then is converged into a thin electron beam focused on the sample surface. A scanning coil is arranged on the edge of the final lens, under which the electron beam is scanned on the sample surface. Due to the interaction between the high-energy electron beam and the sample material, various information are generated: secondary electron, backscattered electron, absorption electron, X-ray, Auger electron, cathodoluminescence and transmission electron, etc. These signals are received by the corresponding receiver, amplified, and sent to the gate of the picture tube to modulate the brightness of the picture tube. Because the current passing through the scanner coil corresponds to the brightness of the picture tube, a bright spot appears on the screen of the picture tube when the electron beam hits the sample bit by bit. The scanning electron microscope (SEM) uses the method of point-by-point imaging to transform the different features of the sample surface into video signal to complete a frame of image, and then various characteristic images of the sample surface is observed on the screen.

5.5.2 Test Result Analysis

The strength and toughness of nano-multiphase ceramics are much higher than those of conventional ceramics. The effect of nano-NiO on the density, the fracture mode, and crack toughening of materials is analyzed by electron microscope.

5.5.2.1 Density

There are larger pores in the samples with nano-NiO content of 0%, the densification process is incomplete, and the sample with nano-NiO content of 10% has smaller pores, as shown in Fig. 5.18. As a result, the relative density of the sample with nano-NiO content of 0% is lower than that of the sample with nano-NiO content of 10%, and the overall mechanical properties of the former one is not as good as that of the latter one. Compared with the spectroscopic analysis in Fig. 5.19, the element Fe accounts for a large component. However, the content of elemental Ni is less, because both NiO and Fe_2O_3 are very stable oxides. Thus, there is no reduction

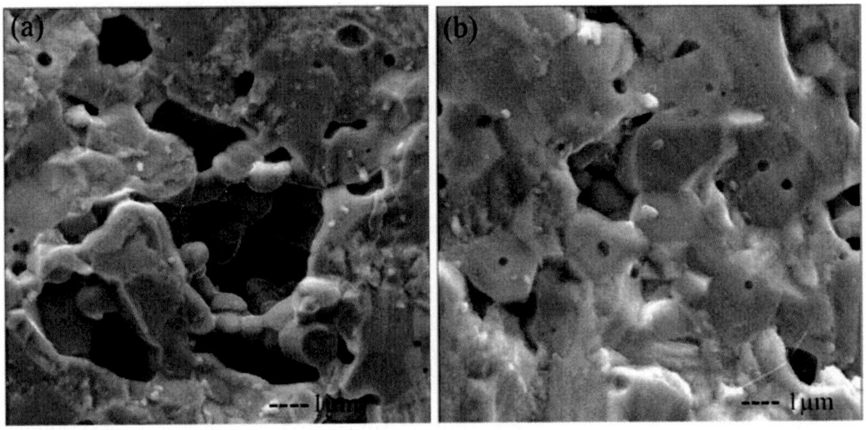

Fig. 5.18 The comparison of structure between samples with nano-NiO contents of 0% and 10%. (**a**) Nano-NiO content is 0%. (**b**) Nano-NiO content is 10%

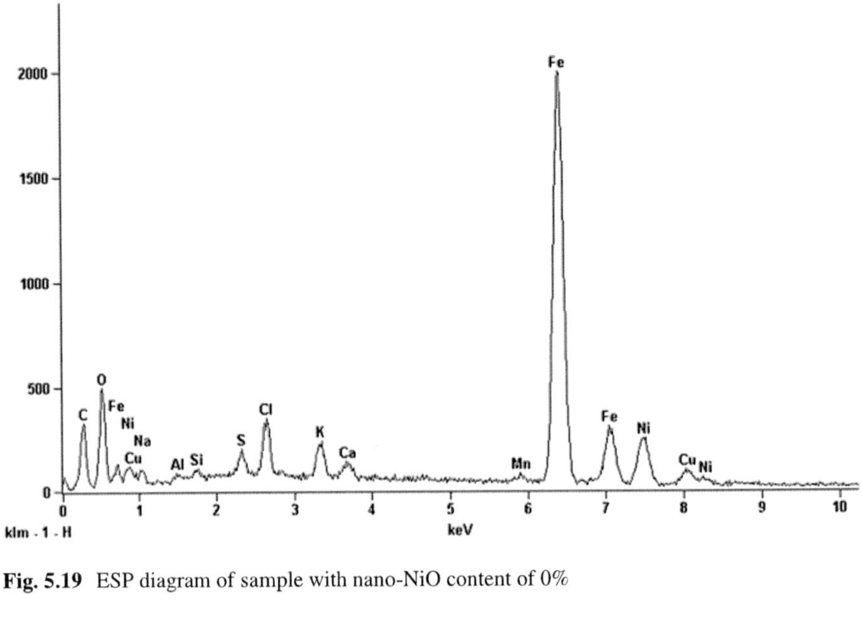

Fig. 5.19 ESP diagram of sample with nano-NiO content of 0%

process in our normal pressure sintering. NiO does not bind well with Fe_2O_3 and has large stomata. The influence of small amount of elements, such as Al, Si, K, and Mn contents on the properties of ceramics can be ignored.

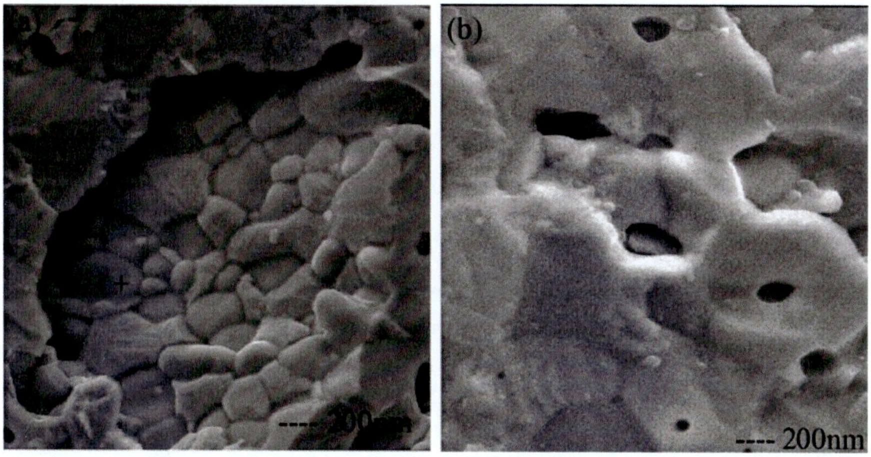

Fig. 5.20 The comparison of fracture surface between samples with nano-NiO contents of 6% and 10%. (**a**) Nano-NiO content of 6% (**b**) Nano-NiO content of 10%

5.5.2.2 Fracture Mode

Corresponding to the analysis of the energy spectrum points in Fig. 5.20, the contents of element Ni and element Fe are roughly the same, indicating that nano-NiO is distributed in the boundary of $NiFe_2O_4$ particles. In Fig. 5.20a, the sample with nano-NiO content of 6% becomes small granular and distributes around the $NiFe_2O_4$. Nano-NiO and matrix $NiFe_2O_4$ materials have good combination, forming intergranular nano-multiphase materials [25]. Nano-NiO has the function of fixing and strengthening grain boundary, which hinders the movement of $NiFe_2O_4$ grain boundary. Thus, the growth of $NiFe_2O_4$ grains during sintering and abnormal growth of $NiFe_2O_4$ grains are inhibited, resulting in the intergranular fracture of the crystals. The mechanical properties such as flexural strength and fracture toughness of sample with nano-NiO are better than those without nano-NiO (Fig. 5.21).

In addition, the nanocrystalline NiO particles on the grain boundary have the effect of pinning and deflecting the crack. During the sintering process, the matrix $NiFe_2O_4$ has a negative effect on the radial compressive stress of the NiO nanoparticles on the grain boundary. This makes the crack propagating along the main grain boundary and "pinning" when it meets the NiO particle. Then, the crack deflects into the crystal to form a small amount of transgranular fracture under the action of greater external force. Therefore, as shown in Fig. 5.20b, samples with nano-NiO content of 10% need to dissipate more energy and exhibit better mechanical properties due to the existence of a small amount of transgranular fracture.

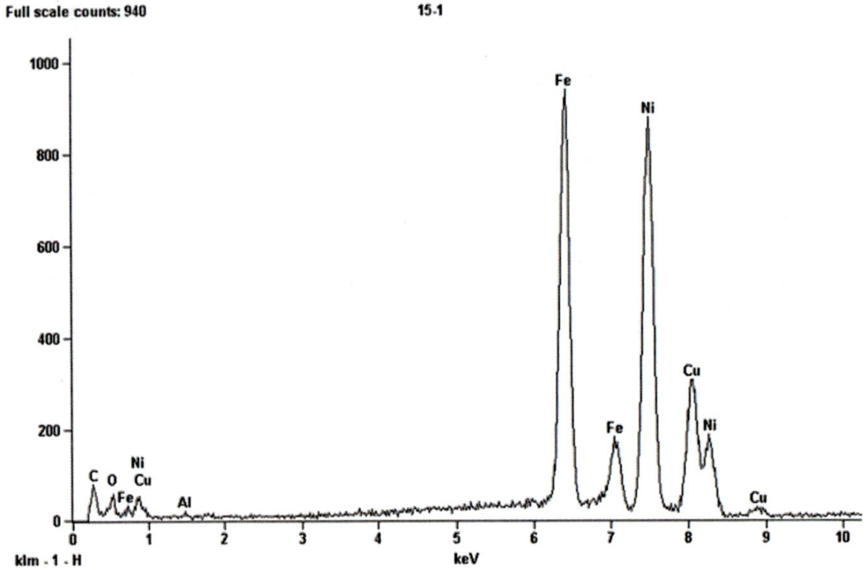

Fig. 5.21 ESP diagram of samples with nano-NiO content of 6%

5.5.2.3 Crack Toughening

As shown in Fig. 5.22a, the crack propagation meets the nano-NiO, and the propagation path is changed. The crack tilts, the non-plane crack surface is formed, the fracture surface increases, and more crack propagation energy is consumed. It plays a toughening role. When cracks propagate through the NiO, the mechanism of crack bridging and toughening begins. As shown in Fig. 5.22b, the NiO particles act as a bridge, the NiO particles are rigid materials, and the strength of the particles is very high. The fracture resistance increases rapidly when a crack bypasses the side of a NiO particle. In addition, the rapid increase of the fracture resistance directly leads to the high fracture strength. In addition, it shows that the flexural strength and fracture toughness of the sample with NiO content of 10% are better than that of the sample with NiO content of 6%.

Compared with Fig. 5.22c without adding nano-NiO, there is a narrow crack in the (c) matrix, and there is no ductile-phase crack bridging between the cracks. This kind of crack produces stress concentration at the tip of the crack under stress, and the propagation is very fast. Therefore, the fracture toughness of the sample without adding nano-NiO is poor.

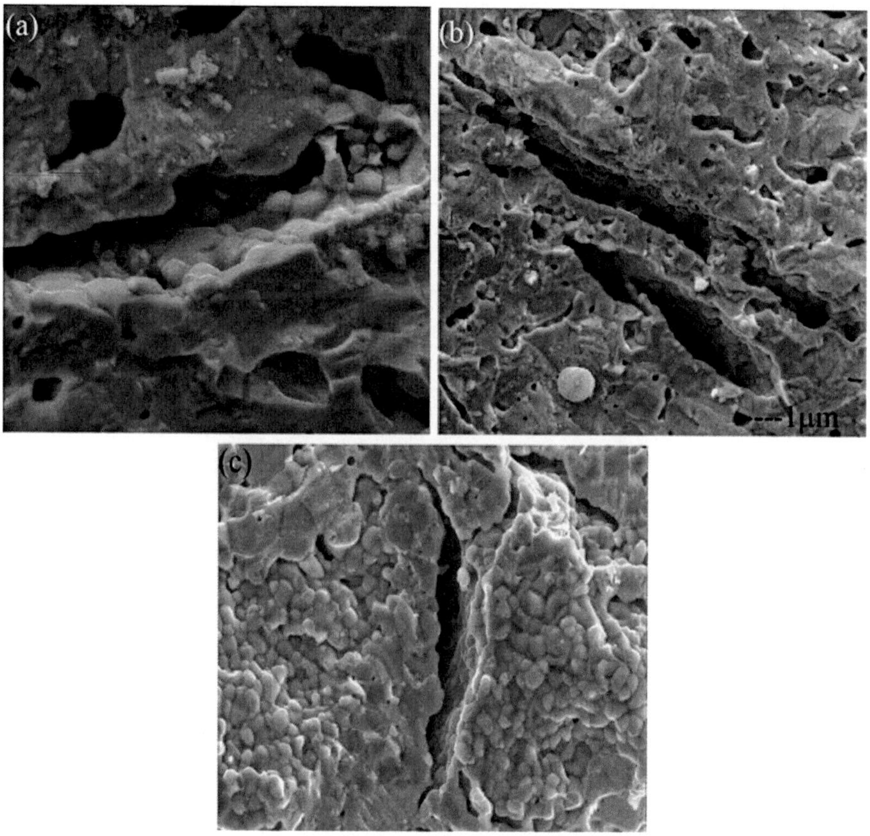

Fig. 5.22 The comparison of crack among samples with nano-NiO contents of 6%, 10%, and 0%. (**a**) Nano-NiO content of 6% (**b**) Nano-NiO content of 10%. (**c**) Nano-NiO content of 0%

5.6 Conclusions

In this paper, the preparation process and mechanical properties of new inert anode for aluminum electrolysis are studied. Nano-sized $NiFe_2O_4$-based cermet samples were prepared based on the proportion of composition and preparation process determined by previous experiments. The structure and mechanical properties of nano-$NiFe_2O_4$-based cermet were characterized, and the following conclusions were obtained.

XRD test shows that the compositions of nano-sized $NiFe_2O_4$ cermet samples are $NiFe_2O_4$, NiO, $CuFe_2O_4$, and CuO. The calculated results show that the sample is nano-$NiFe_2O_4$ cermet.

By using Archimedes drainage method, when the content of nano-NiO is determined at 10%, the density of the sample reaches the maximum value of 95.37, and the sample satisfies the characteristic of high density as an inert anode.

The Vickers hardness was determined at a pressure of 1 kgf and a holding time of 15 s. The Vickers hardness of $NiFe_2O_4$ cermet specimens with nano-NiO of 10% is 476.

The flexural strength of nanocrystalline samples was measured by three-point flexural strength method. The loading rate was set to 0.05 mm/min, and the flexural strength of the samples increases with the increase of nano-NiO content at the beginning. When the content of nano-NiO is 10%, the maximum value is 142.41 MPa. The results show that the addition of nano-NiO can improve the flexural strength of the material.

By scanning electron microscope (SEM), we observed that there is a large porosity in the sample without the addition of nano-NiO. This results in that the density of the sample without NiO is lower than that with the addition of nano-NiO. The nano-NiO can promote the sintering of the sample. The nanocrystalline NiO has a good combination with matrix $NiFe_2O_4$, which forms nano-multiphase material. This inhibits the growth of $NiFe_2O_4$ grain during sintering and $NiFe_2O_4$ grain growth, resulting in the intergranular fracture of $NiFe_2O_4$ crystal. In addition, the crack deflection mode and bridging mechanism exist in crack toughening. Thus, the mechanical properties, such as flexural strength and fracture toughness, of samples with nano-NiO are better than those without nano-NiO.

Finally, we added nano-NiO to cermet Cu-$NiFe_2O_4$ to prepare nano-cermet NiO-Cu-$NiFe_2O_4$. The results show that the corrosion resistance and conductivity of nano-cermet in molten aluminum electrolyte basically meet the requirements of inert anode of aluminum electrolysis. The density, hardness, bending strength, and fracture toughness of the materials have been greatly improved. It can improve the machinability and size of cermet materials.

References

1. Xianjin, Y., Zhuxian, Q., Songzhe, J.: Corrosion of Zinc ferrite in NaF-AlF_3-Al_2O_3 molten salts. J. Chin. Soc. Corros. Protect. **20**(5), 275–279 (2000)
2. Shutin, Z., Guangchun, Y., Yihan, L.: Study on the particle ratio of $NiFe_2O_4$ based inert anode powder. Light Met. **6**, 28–30 (2004) (in Chinese)
3. Yanping, Z.: Physical and Chemical Basis of Electronic Ceramics, pp. 352–383. Electronic Industry Publishing House, Beijing (1996)
4. Japanese Standard Association: Investigation and Research Report on International Standardization of Advanced Resin Matrix Composites, p. 3. Composite Material Association, Tokyo (1993)
5. Zhou, Y.: Science of Ceramic Material, pp. 326–371. Harbin University of Technology Press, Harbin (1995)
6. Marschall, C.W., Rudnick, A.: Fracture Mechanics of Ceramics, vol. 1, p. 301. Plenum Press, New York, NY (1974)
7. Delin, Z.: Powder Metallurgy Material, pp. 152–200. Metallurgical Industry Publishing House, Beijing (1989)
8. Guangyong, L., Tinquan, L., Yu, Z.: Methods for evaluating fracture toughness of ceramic materials. Aerosp. Mat. Tech. **4**, 12–19 (1995) (in Chinese)

9. Jianghong, G.: Fracture Mechanics of Ceramic Materials, pp. 49–64. Tsinghua University Press, Beijing (2001)
10. Tinwei, G., Minwei, P., Bingrong, Z.: The study of the fracture toughness and testing methods of ceramic cutting tool materials. Mat. Mech. Eng. **20**(2), 15–17 (1996) 48. (in Chinese)
11. Gogotsi, G.A.: Fracture toughness of ceramics and ceramics composites. Ceram. Int. **29**, 777–784 (2003)
12. Niihara, K., Morena, R., Hasselman, D.P.H.: Evaluation of K_{1C} of brittle solids by the indentation method with low crack-to-indent ratios. J. Mater. Sci. Lett. **1**, 13–16 (1982)
13. Evans, A.G., Charless, E.A.: Fracture toughness determinations by indentation. Am. Ceram. Soc. **59**(7–8), 371–374 (1976)
14. Lawn, B.R., Marshall, B.: Hardness, toughness and brittleness: an indentation analysis. J. Am. Ceram. Soc. **62**(7–8), 347–350 (1979)
15. Jianghong, G., Zhenduo, G.: Statistical properties of indentation toughness of ceramics. J. Inorg. Mater. **17**(1), 96–104 (2002) (in Chinese)
16. Zener, C.: As communicated by Smith C S. Trans. Am. Inst. Min. Metallurg. Eng. **175**, 48–50 (1948)
17. Niihara, K., Nakahira, A., Tanaka, T.: Nanophase and nanocomposite materials. In: Komarneni, S., Parker, J.C., Thomas, G.J. (eds.) Mater Res Soc Symp Proc, vol. 286, p. 405 (1992)
18. Moya, J.S., Osendi, M.L.: Microstructure and mechanical properties of mullite/ZrO_2 composite. J. Mater. Sci. **19**, 2909–2914 (1984)
19. 新原皓一.: セラミックス **25**(1), 17 (1990)
20. Faber, K.T., Evans, A.G.: Crack deflection process I. Acta Metall. **31**(4), 565–576 (1983)
21. Krstic, V.V., Nicholson, P.S., Hoagland, R.G.: Toughening of glass by metallic particles. J. Am. Ceram. Soc. **64**(9), 499–504 (1981)
22. Marshall, D.B., Morris, W.L., Cox, B.N., Badkhah, M.S.: Toughening mechanisms in cemented carbides. J. Am. Ceram. Soc. **73**(10), 2938–2943 (1990)
23. Ashby, A.E., Blunt, F.J., Bannister, M.: Acta Metall. **37**, 3001–3009 (1989)
24. Tuan, W.H., Brook, R.J.: The toughening of alumina with nickel inclusion. J. Euro. Ceram. Soc. **6**, 31–37 (1990)
25. Niihara, K.: New design concept of Structural ceramics-ceramics nanocomposites. J. Ceram. Soc. Jpn. **99**, 974 (1991)

Index

W. Xianxi, *Inert Anodes for Aluminum Electrolysis*, The Minerals,
Metals & Materials Series, https://doi.org/10.1007/978-3-030-28913-3

D

E

K